# Statistical Data Fusion

# Statistical Data Fusion

Benjamin Kedem
University of Maryland, College Park, USA

Victor De Oliveira
University of Texas at San Antonio, USA

Michael Sverchkov

World Scientific

NEW JERSEY • LONDON • SINGAPORE • BEIJING • SHANGHAI • HONG KONG • TAIPEI • CHENNAI • TOKYO

*Published by*

World Scientific Publishing Co. Pte. Ltd.
5 Toh Tuck Link, Singapore 596224
*USA office:* 27 Warren Street, Suite 401-402, Hackensack, NJ 07601
*UK office:* 57 Shelton Street, Covent Garden, London WC2H 9HE

**Library of Congress Cataloging-in-Publication Data**
Names: Kedem, Benjamin, 1944– | De Oliveira, Victor | Sverchkov, Michael.
Title: Statistical data fusion / by Benjamin Kedem (University of Maryland, College Park, USA), Victor De Oliveira (University of Texas at San Antonio, USA), Michael Sverchkov.
Description: New Jersey : World Scientific, 2017. | Includes bibliographical references and index.
Identifiers: LCCN 2016047612 | ISBN 9789813200180 (hardcover : alk. paper)
Subjects: LCSH: Mathematical statistics. | Mathematical analysis.
Classification: LCC QA276 .K2447 2017 | DDC 519.5--dc23
LC record available at https://lccn.loc.gov/2016047612

**British Library Cataloguing-in-Publication Data**
A catalogue record for this book is available from the British Library.

*To Carmella*
Benjamin Kedem

*To my parents and children*
Victor De Oliveira

*To my family*
Michael Sverchkov

# Preface

There is no single definition of "data fusion," but regardless of one's view, the underlying idea is to come up with estimates or decisions based on multiple data sources as opposed to more narrowly defined estimates or decisions based on single data sources. And as the world is awash with data obtained from numerous and varied processes, there is a need for appropriate statistical methods which in general produce improved inference by taking as input the information from many sources. Motivated by that, this book deals with some aspects of inference based on the combination, or fusion, or integration of many samples, each obtained from (in general) an unknown probability distribution.

The first part of this book, Chapters 1 to 5, deals with semiparametric inference from fused data about probability distributions. The simplest example is when there are only two samples from two different distributions, and the problem is to obtain statistical inference about the two distributions. Clearly, it is possible to estimate each distribution from its own sample. However, under some assumptions, it is possible to obtain improved inference from the combined or fused sample, where each distribution is estimated not from its own sample but from the entire combined data consisting of two samples. The extension to more than two samples, univariate or multivariate, is straightforward. Thus, if there is a network or system of sensors, each producing data according to some probability distribution, the methods discussed in the book deal with the estimation of all the distributions (aka infinite dimensional parameters), as well as the estimation of the finite dimensional parametric connections between them, from the *combined* or *fused* much larger sample. The parametric connection between distributions allows various hypothesis tests including that of equidistribution, that is, the hypothesis that the distributions are identical. In terms of sensors, equidistribution means that all the sensors behave alike. This paradigm of semiparametric estimation from combined data permeates statistics as it will become apparent from the many applications discussed in the book.

Real data can be fused with other real data, or with artificial "fake" data. Thus, a given sample can be fused with computer-generated data giving rise to the notions of *out of sample fusion* and *repeated out of sample fusion*, discussed and applied in Chapter 5.

The main inferential tool used in Chapters 1–5 is the *empirical likelihood* applied to various *density ratio models*. Chapter 6, however, describes a Bayesian extension to tackle some of the same problems considered earlier, where subjective or context-based prior information is incorporated in the analysis. The novel Bayesian analysis that is proposed includes the formulation of prior distributions and a Markov chain Monte Carlo algorithm, tailored to the particulars of the basic density ratio model.

An example of data fusion where the idea is to "borrow strength" from different samples is the important problem of *small area estimation* in the context of informative sampling and nonresponse, discussed in Chapter 7. Here borrowing of strength means using the entire data from areas where information *is* available.

The book has been written with the data practitioner in mind. For that reason we offer examples and illustrations of a wide range of possible applications, from a novel approach to time series prediction to the estimation of small tail probabilities to the estimation of mean body mass index in United States counties. However, there are some parts with enough material to satisfy the more mathematically inclined.

**Acknowledgment**: In writing this book, the works of several of the first author's former graduate students, all mentioned in the bibliography, were instrumental. He is grateful to each and every one of them. In particular he would like to thank Kevin Dayaratna, Konstantinos Fokianos, Richard Gagnon, Haiming Guo, Guanhua Lu, Lemeng Pan, Anastasia Voulgaraki, Shihua Wen, and Wen Zhou.

# Contents

# Chapter 1

# Introduction

*"Give me a place to stand and rest my lever on, and I can move the Earth."* (Archimedes, 287-212 B.C.)

A great deal of the statistical literature deals with a single sample coming from a distribution, univariate or multivariate, and the problem is to identify the distribution or parts thereof by an array of estimation and testing procedures. As such, this practice neglects to bring in information from other sources which could improve the desired inference. Similarly, given time series records, the analysis quite often is based on a single time series, not considering many other related time series to bear upon the problem of interest such as prediction.

On numerous occasions, however, data from many sources are available to the statistician. Examples include case-control data, numerous related economic time series, weather measurements from different instruments, data from factorial designs, data from many sensors in a surveillance system, microarray data when considering each gene as a data source, and even computer generated artificial data if incorporated or fused wisely with real data sources—we shall see examples of this in the estimation of tail probabilities. This observation is the driving force behind this book which is about combination or fusion of statistical information from multiple sources. We shall refer to this interchangeably as data or information fusion, combination, or integration.

We can make the case of why information fusion is useful, and in many cases even necessary, by bringing forward an analogy with the way living organisms process and internalize information. Thus, imagine an organism which relies on a single sense for its survival, for example the sense of thermoception only. Chances are it will have a hard time competing against other organisms which have the ability to sense and process information from multiple sources using multiple faculties. Arguably, the latter is a more efficient way to perceive and learn about the ambient world around these organisms. A case in point is that humans' retention of what is said is much higher when the verbal information is coupled with graphical or pictorial information. This is entirely analogous to a highly specialized medical treatment versus a more holistic sensical approach to medical care, based on data from multiple medical tests concerning more than just the "left toe" as it were. And there is clearly a direct analogy here with statistical inference based on a single data source versus one which is based on many sources.

Another important case for data integration has to do with "small area" problems where samples within designated domains, such as a school district or county, are too small for any meaningful inference which necessitates "borrowing of strength" from additional related data sources for more precise estimation.

As we shall see, approaching statistical inference with the view of data integration in mind, sheds new light on some well known statistical procedures. For example, against the background of information fusion, both regression and analysis of variance can be reinterpreted and approached afresh in a novel way, bypassing linearity and the normal assumption. In general, information fusion leads to modified and mostly more precise statistical procedures including case-control studies, logistic regression, kernel density estimation, finite population surveys, and time series prediction. We can add to this partial list other statistical procedures which can be improved substantially when the analysis is based on data combined from multiple samples or sources. A key idea behind all this is the notion of a system of distributions, representing multiple data sources, of which one distribution serves as a *reference* distribution and the rest are *distortions* or deviations from the reference distribution. An important fact is that under a single assumption we can get useful analytical approximations for these distributions, be they univariate or multivariate, without recourse to the normal assumption.

## 1.1 A Reference Distribution and its Distortions

The notion of a reference distribution and its many distortions can be motivated from a general problem in remote sensing, such as remote sensing from space. Thus, consider a collection of satellite borne sensors $I_1, ..., I_q$ remotely sensing a geophysical quantity from a great distance away in space, where due to the distance the measurements are biased or "tilted" relative to ground truth. Suppose the data from instrument $I_j$ are distributed according to a probability density function (pdf) $g_j$, $j = 1, ..., q$, and let $g_{q+1} \equiv g_m \equiv g$ be the pdf representing ground truth. Then it is reasonable to entertain a model of the type,

$$\begin{aligned} g_1 &= w_1 g \\ &\cdot \\ &\cdot \\ &\cdot \\ g_q &= w_q g \end{aligned} \tag{1.1}$$

where the $w_i$ are appropriate nonnegative weights. That is, each $g_j$ is a weighted *distortion* relative to the *reference* $g$. As such, the system (1.1) is a general *model*, special cases of which are *widespread throughout statistics* as will be illustrated next.

Rao (1965) and Patil and Rao (1977) refer to the $g_j(x)$ as *weighted distributions*. In their notation, if $g(x)$ denotes the pdf of a random variable $X$, then the weighted distribution associated with a nonnegative function $w(x)$ is the distribution of the random variable $X^w$ with pdf

$$g^w(x) = \frac{w(x)g(x)}{E(w(X))}, \tag{1.2}$$

provided that $E(w(X))$ exists. This type of distribution arises often in situations in which there are severe limitations in the sampling or measuring process. Suppose that a random mechanism produces a value $x$ according to the pdf $g$, but such value may or may not be observed, depending on some random mechanism. If $w(x) = P(\text{observation is made} \mid X = x)$, then (1.2) provides the distribution of the observed values. Although this interpretation on the genesis of weighted distributions holds in many situations, (1.2) serves as a sensible model in many other situations in which $w(x)$ is not necessarily a probability (so it may take values larger than 1). Here are two common examples.

**Truncated distributions:** Let $g(x)$ be a pdf and $T$ a subset of its support. Suppose that values generated from $g(x)$ can be observed (and hence enter into a sample) only when they belong to $T$. In this case the distribution of the observed values is the weighted distribution (1.2) with $w(x) = I(x \in T)$, where $I(B)$ denotes the indicator function of event $B$.

**Length-biased sampling:** Suppose the size of rocks of a certain kind in an area follows the distribution $g(x)$. If the data collection process is such that large rocks are more likely to be sampled than small rocks (because they are easier to spot), then a sensible model for the distribution of the sampled rocks size is the weighted distribution (1.2) with $w(x) = x$.

Vardi (1982) studied the nonparametric estimation of the distribution functions $F, G$ in the *length-biased sampling* model

$$F(y) = \frac{1}{\mu}\int_0^y x dG(x),\ y \geq 0,$$

where $\mu = \int_0^\infty x dG(x) < \infty$. Later, Vardi (1985), and Gill, Vardi, Wellner (1988) considered the more general biased sampling model

$$F(y) = \frac{1}{W(G)}\int_{-\infty}^y w(x) dG(x),$$

where $w(x)$ is known and $W(G) = \int_{-\infty}^\infty w(x)dG(x) < \infty$. A further extension is discussed in Gilbert et al (1999). Differentiation shows that both biased sampling models are special cases of (1.3) below with $q = 1$.

Unlike the previous examples in which the weight function is known, in most cases the weight function is just partialy specified as it would depend on unknown parameters. An important special case of (1.1) is obtained when each probability density $g_j(x)$, $j = 1, ..., q$, is an exponential tilt or distortion of a reference $g_m(x)$,

$$\frac{g_j(x)}{g_m(x)} = \exp(\alpha_j + \boldsymbol{\beta}_j' \boldsymbol{h}(x)),\ j = 1, \ldots, q \tag{1.3}$$

where $\boldsymbol{\beta}_j$ is a vector parameter and $\boldsymbol{h}(x)$ is an arbitrary but known vector-valued function of $x$. In the context and notation of (1.2), $g_j(x)$ is the weighted distribution obtained from the pdf $g_m(x)$ and the weight function $w_j(x) = \exp(\boldsymbol{\beta}_j' \boldsymbol{h}(x))$, in which case $E_{g_m}(w_j(X)) = \exp(-\alpha_j)$. Nevertheless, for most of this book we view the ratio $w(x)/E_{g_m}(w(X))$ as the weight or distortion function and denote it as $w(x)$. For each $j$, and also collectively for all $j$, (1.3) is referred to as a *density ratio model.* Taking logarithms of (1.3)

$$\log\left\{\frac{g_j(x)}{g_m(x)}\right\} = \alpha_j + \boldsymbol{\beta}_j'\boldsymbol{h}(x),\ j = 1,\ldots,q \tag{1.4}$$

gives $q$ linear models, however, when the $\alpha_j + \boldsymbol{\beta}_j'\boldsymbol{h}(x)$ are replaced by nonlinear expressions the logarithms result in $q$ nonlinear models. Observe that the density ratio model is reminiscent of the celebrated Cox proportional hazards model with respect to a baseline (reference) hazard.

Density ratio models have been with us for a long time and there is a sizable body of literature dealing with various types of density ratios models and their applications. They arise particularly in classification and discrimination problems in discrete and continuous data, and in case-control studies, also known as retrospective studies, where the problem is to compare a sample (at times different samples) of disease carriers (cases) to a sample of disease-free individuals (controls). Early references include Wald (1944), Anderson (1972, 1979), Prentice and Pyke (1979), and Cox and Snell (1989), among others. Several examples will drive home this point.

**Multinomial logistic regression:** To see how multinomial logistic regression in particular morphs into a density ratio model as in (1.3), consider a categorical random variable $y$ such that $P(y=j) = \pi_j$, $\sum_{j=1}^m \pi_j = 1$, and $f(x \mid y=j) = g_j(x)$, $j = 1, ..., m$, with $m = q+1$. If for $j = 1, ..., m$

$$P(y = j \mid x) = \frac{\exp(\alpha_j^* + \beta_j h(x))}{1 + \sum_{k=1}^q \exp(\alpha_k^* + \beta_k h(x))},$$

then an appeal to Bayes theorem, using the unspecified marginal distribution of $x$, shows that (1.3) holds with $\alpha_j = \alpha_j^* + \log(\pi_m/\pi_j)$, $j = 1, ..., q$, and $\alpha_m^* = 0, \beta_m = 0$.

A multivariate version of (1.3) is obtained within the case-control framework studied in Prentice and Pyke (1979). Let $D = i$ denote the $j$th disease incidence, $j = 1,\ldots,q$, and let $D = m$ indicate disease-free state. With $\pi_j = P(D=j)$, $\sum_{j=1}^m \pi_j = 1$, let $P(D = j \mid \mathbf{x})$ denote the conditional probability that an individual with covariate vector $\mathbf{x}$ has disease $D = i$, where $\mathbf{x} \sim f(\mathbf{x})$. Define $g_j(\mathbf{x}) = f(\mathbf{x} \mid D = j)$, $j = 1,\ldots,q,m$, and assume the generalized logistic regression model,

$$P(D = j \mid \mathbf{x}) = \frac{\exp(\alpha_j^* + \boldsymbol{\beta}_j'\mathbf{x})}{1 + \sum_{k=1}^q \exp(\alpha_k^* + \boldsymbol{\beta}_k'\mathbf{x})},\quad j = 1, ..., q, m \tag{1.5}$$

where $\alpha_m^* = 0$ and $\boldsymbol{\beta}_m = \mathbf{0}$ for (1.5) to be well defined. Then, from Bayes theorem we obtain the density ratio model

$$\frac{g_j(\mathbf{x})}{g_m(\mathbf{x})} = \exp(\alpha_j + \boldsymbol{\beta}_j'\mathbf{x}), \quad j = 1, \ldots, q \tag{1.6}$$

where $\alpha_j = \alpha_j^* + \log(\pi_m/\pi_j)$. Holding $g \equiv g_m$ as a reference density we obtain (1.1).

The connection with logistic regression points to the fact that the $\boldsymbol{\beta}_j$ can be estimated assuming a model such as (1.5), which provides only part of the puzzle regarding the infinite dimensional parameters $g_j(\mathbf{x})$. On the other hand, as we shall see, the density ratio model (1.6) allows direct semi-parametric inference about both the $\boldsymbol{\beta}_j$ and the multivariate probability densities $g_j(\mathbf{x})$. Moreover, (1.6) plays on the intuitively appealing notion of a baseline behavior, or reference, and deviations from it.

Disregarding their logistic regression origin, (1.3) and (1.6) may be viewed as "regression" models connecting probability distributions, which, interestingly enough, can contort back into logistic regression by conditioning on appropriate categorical random variables in conjunction with Bayes rule.

**Classical one-way ANOVA:** Consider the classical one-way ANOVA with $m = q + 1$ independent normal random samples, where the $j$th sample has $n_j$ observations from $\mathrm{N}(\mu_j, \sigma^2)$ with probability density function (pdf) $g_j(x)$. In symbols, for $j = 1, \ldots, q, q + 1 = m$,

$$\{x_{j1}, ..., x_{jn_j}\} \sim g_j(x) \sim \mathrm{N}(\mu_j, \sigma^2). \tag{1.7}$$

Then, holding $g_m(x) \equiv g(x)$ as a reference, the density ratio model (1.3) emerges:

$$\begin{aligned} g_1(x) &= \exp\{\alpha_1 + \beta_1 x\} g(x) \\ &\cdot \\ &\cdot \\ &\cdot \\ g_q(x) &= \exp\{\alpha_q + \beta_q x\} g(x) \end{aligned} \tag{1.8}$$

with

$$\alpha_j = \frac{\mu_m^2 - \mu_j^2}{2\sigma^2}, \quad \beta_j = \frac{\mu_j - \mu_m}{\sigma^2}, \quad j = 1, \ldots, q.$$

If the variances are not the same and $g_j(x)$ is the pdf of $\mathrm{N}(\mu_j, \sigma_j^2)$, then again the same structure persists but with different weights,

$$\begin{aligned} g_1(x) &= \exp\{\alpha_1 + \beta_1 x + \gamma_1 x^2\} g(x) \\ &\cdot \\ &\cdot \\ &\cdot \\ g_q(x) &= \exp\{\alpha_q + \beta_q x + \gamma_q x^2\} g(x) \end{aligned} \tag{1.9}$$

where for $j = 1, \ldots, q$

$$\alpha_j = \log\left(\frac{\sigma_m}{\sigma_j}\right) + \frac{\mu_m^2}{2\sigma_m^2} - \frac{\mu_j^2}{2\sigma_j^2}, \ \beta_j = \frac{\mu_j \sigma_m^2 - \mu_m \sigma_j^2}{\sigma_j^2 \sigma_m^2}, \ \gamma_j = \frac{\sigma_j^2 - \sigma_m^2}{2\sigma_j^2 \sigma_m^2} \tag{1.10}$$

and $\mu_m, \sigma_m^2$ are the parameters of the reference pdf $g \equiv g_m$.

**Bernoulli samples:** If the data in (1.7) are binary such that

$$x_{j1}, \ldots, x_{jn_j} \sim \mathrm{Bernoulli}(\pi_j)$$

then a weighted system identical to (1.8) holds with

$$\alpha_j = \log\left(\frac{1-\pi_j}{1-\pi_m}\right), \quad \beta_j = \log\left(\frac{\pi_j(1-\pi_m)}{\pi_m(1-\pi_j)}\right), \quad j = 1, ..., q.$$

**Tilting the uniform distribution:** Let $g \equiv g_2$ be the pdf of the uniform distribution on $(0,1)$, and truncate an exponential distribution to $(0,1)$ as $g_1(x) = \exp(-x)/(1-\exp(-1))$, $x \in (0,1)$. Then with $\alpha = -\log(1 - \exp(-1)), \beta = -1$,

$$g_1(x) = \exp(\alpha + \beta x) g(x)$$

a special case of (1.1) with $q = 1$.

**Distributions known up to normalizing constants:** Fokianos and Qin (2008) consider $q$ independent random samples from the probability densities

$$g_j(x, \boldsymbol{\theta}_j) = \frac{p_j(x, \boldsymbol{\theta}_j)}{c_j(\boldsymbol{\theta}_j)}, \quad j = 1, ..., q, \tag{1.11}$$

where $p_j(x, \boldsymbol{\theta}_j)$ is a known non-negative function, $\boldsymbol{\theta}_j$ is a $p$-dimensional parameter and $c_j(\boldsymbol{\theta}_j) = \int p_j(x, \boldsymbol{\theta}_j) dx$ is a normalizing constant. Assume the support of $g_j$ is independent of $\boldsymbol{\theta}_j$, and let $g(x)$ be a probability density over the same support as that of $g_j$. Then, with obvious notation,

$$\begin{aligned} g_j(x, \boldsymbol{\theta}_j) &= \frac{p_j(x, \boldsymbol{\theta}_j)}{c_j(\boldsymbol{\theta}_j)} \frac{g(x)}{g(x)} \\ &= \exp\{-\log c_j(\boldsymbol{\theta}_j) + (\log p_j(x, \boldsymbol{\theta}_j) - \log g(x))\} g(x) \\ &\equiv \exp(\alpha_j + \phi_j(x; \boldsymbol{\theta}_j))g(x). \end{aligned} \tag{1.12}$$

With $j = 1, ..., q$, (1.12) is a special case of (1.1). This construction is used in connection with importance sampling in the estimation of the normalizing constants, where random samples are available from each $g_j$ but artificial data are generated from the reference $g$.

There are situations that deal with a *single* dataset in which the above modeling framework is useful, either because the models display an internal structure that can be brought to match (1.1) or a formal analogy is present. The following examples illustrate this.

**K-Parameter exponential families:** A great generalization of the previous normal and Bernoulli cases is achieved by considering the general k-parameter exponential family

$$g(x, \boldsymbol{\theta}) = d(\boldsymbol{\theta})S(x) \exp\left\{\sum_{i=1}^{k} c_i(\boldsymbol{\theta})T_i(x)\right\}$$

where $\boldsymbol{\theta} = (\theta_1, ..., \theta_k)$. To see the structure (1.3) emerging, it is sufficient to consider two distinct values of $\boldsymbol{\theta}$. Then, with $\alpha = \log[d(\boldsymbol{\theta}_1)/d(\boldsymbol{\theta}_2)]$, $\boldsymbol{\beta} = (c_1(\boldsymbol{\theta}_1) - c_1(\boldsymbol{\theta}_2), ..., c_k(\boldsymbol{\theta}_1) - c_k(\boldsymbol{\theta}_2))'$, and $\boldsymbol{h}(x) = (T_1(x), ..., T_k(x))'$, we obtain the ratio

$$\frac{g_1(x)}{g_2(x)} \equiv \frac{g(x, \boldsymbol{\theta}_1)}{g(x, \boldsymbol{\theta}_2)} = \exp\{\alpha + \boldsymbol{\beta}'\boldsymbol{h}(x)\} \tag{1.13}$$

or

$$g_1(x) = \exp\{\alpha + \boldsymbol{\beta}'\boldsymbol{h}(x)\}g_2(x). \tag{1.14}$$

In the normal case with mean $\mu$ and variance $\sigma^2$, $\boldsymbol{\theta} = (\mu, \sigma^2)$, we obtain $\alpha, \boldsymbol{\beta}$ from (1.10). In particular, from (1.9)

$$\boldsymbol{h}(x) = (x, x^2)'. \tag{1.15}$$

The corresponding lognormal distribution with $\boldsymbol{\theta} = (\mu, \sigma^2)$ gives

$$\boldsymbol{h}(x) = (\log x, \log^2 x)', \tag{1.16}$$

which reduces to $h(x) = \log x$ when $\sigma_1^2 = \sigma_2^2$. In the gamma case with shape $r$ and scale $\lambda$, $\boldsymbol{\theta} = (r, \lambda)$,

$$\alpha = \log \frac{\lambda_1^{r_1}\Gamma(r_2)}{\lambda_2^{r_2}\Gamma(r_1)}, \quad \boldsymbol{\beta} = (\lambda_2 - \lambda_1, r_1 - r_2)', \quad \boldsymbol{h}(x) = (x, \log x)', \tag{1.17}$$

which reduces to $h(x) = \log x$ when $\lambda_1 = \lambda_2$. As for the Rayleigh distribution with scalar parameter $\theta$, (1.14) holds with

$$\alpha = \log \frac{\theta_2^2}{\theta_1^2}, \quad \beta = \frac{1}{2\theta_2^2} - \frac{1}{2\theta_1^2}, \quad h(x) = x^2.$$

Kay and Little (1987) list many more special cases of (1.14).

**Comparison densities:** Another example of the general system (1.1) is obtained in terms of *comparison densities*, a concept advanced by Emanuel Parzen in many of his papers, for example Parzen (2004). Accordingly, consider continuous distribution functions $F_1, ..., F_q$ and $G$ with probability densities $f_1, ..., f_q$ and $g$, respectively, and let

$$D(u; G, F_j) = F_j(G^{-1}(u)), \quad 0 < u < 1.$$

The corresponding *comparison densities* are defined as

$$d(u; G, F_j) = f_j(G^{-1}(u))/g(G^{-1}(u))$$

or

$$f_j(G^{-1}(u)) = d(u; G, F_j)g(G^{-1}(u)), \quad j = 1, ..., q \tag{1.18}$$

which has the same structure as (1.1). An analogous relationship holds in the discrete case. Comparison densities are useful for assessing goodness of fit and equality of distributions, since $F_1 = \ldots = F_q = G$ if and only if $d(u; G, F_j) = 1$ for all $u$ in $(0, 1)$ and $j = 1, ..., q$ (Thas, 2010).

**Prior-Posterior relationships:** Another formal illustration of weighted distributions was provided by Patil, Rao and Ratnaparkhi (1986) in the context of a Bayesian framework. Let $\boldsymbol{y} = (y_i, ..., y_q)'$ be a random vector of observables with joint distribution $\bar{p}(\boldsymbol{y}|\boldsymbol{\theta})$, where $\boldsymbol{\theta}$ are unknown parameters having prior distribution $\pi(\boldsymbol{\theta})$. Then the posterior distribution of $\boldsymbol{\theta}$ is

$$\begin{aligned}\pi(\boldsymbol{\theta}|\boldsymbol{y}) &= \frac{\bar{p}(\boldsymbol{y}|\boldsymbol{\theta})\pi(\boldsymbol{\theta})}{\int \bar{p}(\boldsymbol{y}|\boldsymbol{\xi})\pi(\boldsymbol{\xi})d\boldsymbol{\xi}} \\ &= \frac{L(\boldsymbol{\theta};\boldsymbol{y})\pi(\boldsymbol{\theta})}{E_\pi(L(\boldsymbol{\Theta};\boldsymbol{y}))},\end{aligned}$$

where $L(\boldsymbol{\theta};\boldsymbol{y}) = \bar{p}(\boldsymbol{y}|\boldsymbol{\theta})$. Hence the posterior distribution is a weighted distribution, as in (1.2), where the likelihood plays the role of weight function.

Suppose now the observations are identically distributed, with conditional marginal pdf $p(y|\boldsymbol{\theta})$, but their joint distribution is not specified. Then, we would have the $q$ prior-posterior relationships

$$\begin{aligned}\pi(\boldsymbol{\theta}|y_1) &= \frac{p(y_1|\boldsymbol{\theta})}{p(y_1)}\pi(\boldsymbol{\theta}) \\ \pi(\boldsymbol{\theta}|y_2) &= \frac{p(y_2|\boldsymbol{\theta})}{p(y_2)}\pi(\boldsymbol{\theta}) \\ &\cdot \\ &\cdot \\ &\cdot \\ \pi(\boldsymbol{\theta}|y_q) &= \frac{p(y_q|\boldsymbol{\theta})}{p(y_q)}\pi(\boldsymbol{\theta}),\end{aligned} \tag{1.19}$$

where the $p(y_j)$ are the marginal distributions of the observations. Hence the system of posterior distributions based on single observations follows model (1.1), where the prior plays the role of the reference distribution and $w_j(\boldsymbol{\theta}) = p(y_j|\boldsymbol{\theta})/p(y_j)$ for $j = 1, \ldots, q$.

**Spectral analysis:** A formal structure analogous to (1.1) exists in terms of spectral densities. A spectral density of a real-valued stationary process is essentially a probability density up to a normalizing constant. The celebrated relationship between the spectral densities of the input, $f(\omega)$, and of the output, $f_H(\omega)$, in a linear system with gain $|H(\omega)|$, states that (Koopmans 1974, p. 86)

$$f_H(\omega) = |H(\omega)|^2 f(\omega), \quad \omega \in (-\pi, \pi].$$

If the input series is operated on by a sequence of $q$ linear filters with gains $|H_1(\omega)|, ..., |H_q(\omega)|$, then with output spectral densities $f_1(\omega), ..., f_q(\omega)$,

$$\begin{aligned} f_1(\omega) &= |H_1(\omega)|^2 f(\omega) \\ &\cdot \\ &\cdot \\ &\cdot \\ f_q(\omega) &= |H_q(\omega)|^2 f(\omega). \end{aligned} \tag{1.20}$$

That is, multiple "distortions" similar to (1.1) of the same *reference* spectral density $f$. We point to the system (1.20) only because of its intriguing formal structure which transcends "regular" statistics as is evident from the previous examples but do not intend to pursue (1.20) any further.

In summary, a general structure emerges from the previous examples of a reference behavior (distribution) and its many distortions of the form (1.1). Many other examples will be described in detail in the following chapters.

### 1.1.1 A General Problem

We have seen that systems of weighted distributions as in (1.1) are quite ubiquitous and often encountered in statistical applications. This encourages the study of the general structure (1.1), of a reference behavior (distribution) and its many distortions. The question is how to go about doing this. The clue "cries out" to us from, for example, the relationships (1.6) (1.8), (1.9), and (1.14) when data are available from each distribution, reference as well as distortions. These relationships involve both finite dimensional parameters ($\alpha$'s and $\boldsymbol{\beta}$'s) and infinite dimensional parameters in the form of probability densities ($g_i$'s), and hence a *semiparametric* approach is appropriate, and as we shall see this approach is also useful and worthwhile.

Accordingly, the above discussion leads us to the following formulation of a general semiparametric problem to be discussed and applied in the next chapters.

**Fusion:** Given $m = q + 1$ independent random samples $\boldsymbol{x}_1, ..., \boldsymbol{x}_q, \boldsymbol{x}_m$, from multiple data sources, where $\boldsymbol{x}_j = (x_{j1}, \ldots, x_{jn_j})'$ is of size $n_j$, fuse or combine the data in one long vector

$$\boldsymbol{t} = (t_1, ..., t_n)' \equiv (\boldsymbol{x}_1', ..., \boldsymbol{x}_q', \boldsymbol{x}_m')'$$

of length $n \equiv n_1 + \cdots + n_q + n_m$.

**Density ratio model**: Assume $x_{ji} \sim g_j(x)$ for $j = 1, ..., q, m$, $i = 1, ..., n_j$, and let $g_m(x) = g(x)$ be the reference pdf. For a given distortion function $\boldsymbol{h}(x)$, assume the $q = m - 1$ distortions of the reference $g$,

$$g_j(x) = \exp\{\alpha_j + \boldsymbol{\beta}_j' \boldsymbol{h}(x)\} g(x), \qquad j = 1, ..., q. \tag{1.21}$$

**Problems**: Use the combined data $\boldsymbol{t} = (t_1, ..., t_n)'$ to

1. Estimate the reference pdf $g(x)$ and the corresponding cdf $G(x)$.

2. Estimate $\boldsymbol{\alpha} = (\alpha_1, ..., \alpha_q)'$, $\boldsymbol{\beta} = (\boldsymbol{\beta}_1', ..., \boldsymbol{\beta}_q')'$.

3. Test hypotheses about the $\boldsymbol{\beta}_i$, and in particular test distribution equality (equidistribution), $\text{H}_0$: $\boldsymbol{\beta}_1 = \cdots = \boldsymbol{\beta}_{\text{q}} = \mathbf{0}$.

4. Predict future observations.

**Remark.** Here and elsewhere, for the sake of clarity we shall ignore the problem of ties which is less of a problem in continuous samples but exacerbated in discrete data. Attention to this problem is paid in Vardi (1982, 1985), Gilbert et al. (1999), and Owen (2001). Taking ties into account results in a slight likelihood modification which is not essential for our development.

As such, (1.21) is a system of *density ratio models* where various distributions are "regressed" on a common or baseline reference distribution, and the problem is to estimate the reference pdf-CDF pair $g, G$ from the entire combined data $\boldsymbol{t}$ and not just from the corresponding "reference" sample $\boldsymbol{x}_m$. Likewise we wish to estimate $\boldsymbol{\beta}_1$ from the combined or integrated data $\boldsymbol{t}$ and not just from the corresponding samples $\boldsymbol{x}_1$ and $\boldsymbol{x}_m$, and so on. In fact the problem is to "estimate everything from everything". Also, observe that when $\boldsymbol{\beta}_1 = \cdots = \boldsymbol{\beta}_q = \mathbf{0}$ then all the $\alpha_j = 0$ as well, and hence all the $g_j$ are equal to reference $g$.

The preceding semiparametric formulation opens the door to a useful general statistical approach based on *fused information* from many sources. This general construction does not require normality or even symmetry of the distributions, and the density ratio model does not require knowledge of the reference distribution, while accommodating both continuous as well as discrete distributions. The main assumption is the form of the distortion of the reference $g(x)$, namely $\exp\{\alpha_j + \boldsymbol{\beta}_j' \boldsymbol{h}(x)\}$, which requires the choice of a "distortion function" or "tilt function" $\boldsymbol{h}(x)$. That is, the distortion function $\boldsymbol{h}(x)$ is assumed to be known. Experience tells us that the choices $h(x) = x$ and $\boldsymbol{h}(x) = (x, x^2)'$, as in (1.8) and (1.9), are useful for symmetric data, as is $\boldsymbol{h}(x) = (x, \log x)$, or more generally $\boldsymbol{h}(x) = (x^{\tau}, \log x)$, when the data are skewed. As we shall see, in many cases a density ratio model with a misspecified $\boldsymbol{h}(x)$ may still yield useful inference. Finally, notice that the reference distribution may be any of the $m$ distributions in the system, leaving the exponential distortion intact but with shifted parameters.

# Chapter 2

# Weighted Systems of Distributions

*"No finite point has meaning without an infinite reference point."*
(Jean-Paul Sartre, 1905-1980.)

This chapter deals with inference for density ratio models, first for one-dimensional, and then for p-dimensional tilt functions.

The previous chapter showed that the notion of a system of distributions representing multiple data sources, of which one system serves as a reference distribution and the rest are distortions or deviations from the reference distribution, is widespread throughout statistics. In this chapter we examine the model more closely, and provide more details. The model is semiparametric in nature, and as such it does not require specifically the ubiquitous normal assumption, although the useful tilt model (1.15) is inspired by the ratio of two normal density functions. The *empirical likelihood*, a most useful tool in semiparametric inference, is the main inferential tool in this as well as in the next few chapters.

## 2.1 Simple Weighted Systems

The basic structure of one-way ANOVA under the normal assumption displayed in (1.8) fits the paradigm of a weighted system of distributions. A slight generalization of this structure is a good starting point for illustrating our approach to inference for weighted systems.

An immediate generalization is obtained by first eliminating the normal assumption altogether and regarding each $g_j(x)$ directly as an exponential distortion or tilt of the reference $g(x) \equiv g_m(x)$, and second, replacing $x$ in $\exp\{\alpha_i + \beta_i x\}$ by an arbitrary but known function $h(x)$. For example $h(x) = \log x$, or $h(x) = x^2$. Clearly, $h(x) = x$ is always a possibility. Then the weighted system (1.8) upgrades to the more general weighted system

$$g_j(x) = \exp\{\alpha_j + \beta_j h(x)\}g(x), \; j = 1, ..., q. \tag{2.1}$$

Clearly, $\alpha_j$ depends on $\beta_j$. In particular, by integrating both sides of (2.1) we see that, for $h(x) = x$, $\alpha_j$ is determined explicitly by $\beta_j$ through the moment generating function $M_g$ corresponding to $g$,

$$\alpha_j = -\log\{M_g(\beta_j)\}, \; j = 1, ..., q.$$

A similar relation between $\alpha_j$ and $\beta_j$ holds for each choice of $h(x)$.

We shall refer to (2.1) as a *simple weighted system* (SWS). More general systems of weighted distribution will be studies in Section 2.3.

Following the general formulation in Section 1.1.1, a generalization of the classical one-way ANOVA in terms of SWS then runs as follows. Use the combined or fused data from $m$ independent random samples

$$\boldsymbol{t} = (t_1, ..., t_n)' \equiv (\boldsymbol{x}_1', ..., \boldsymbol{x}_q', \boldsymbol{x}_m')'$$

of length $n \equiv n_1 + \cdots + n_q + n_m$, where $\boldsymbol{x}_j = (x_{j1}, ..., x_{jn_j})'$, to estimate the reference pdf $g(x)$, the corresponding cdf $G(x)$, the parameters $\boldsymbol{\alpha} = (\alpha_1, ..., \alpha_q)'$, $\boldsymbol{\beta} = (\beta_1, ..., \beta_q)'$, and study the large sample properties of the estimators. Also, test the equidistribution hypothesis $\text{H}_0 : \beta_1 = \cdots = \beta_q = 0$.

### 2.1.1 Some Properties

In this section we list a few properties of the SWS of distributions (2.1) that illustrate its scope and flexibility.

**Property 1.** Under the SWS of distributions (2.1) the equidistribution hypothesis $g_1 = \cdots = g_q = g_m$ is equivalent to $\text{H}_0 : \beta_1 = \cdots = \beta_{\text{q}} = 0$.

Since $g(x)$ is a density, $\beta_j = 0$ implies $\alpha_j = 0$, $j = 1, ..., q$, and hence the hypothesis $\text{H}_0 : \beta_1 = \cdots = \beta_{\text{q}} = 0$ implies that all the $m$ populations are equidistributed, namely, $g_j = g$ for all $j$. The converse is also clear. Recall that under the classical one-way ANOVA with normally distributed samples with equal variance, testing for mean equality is really a test about equidistribution.

**Property 2.** Consider the SWS of distributions (2.1) with pdfs $g_1, \ldots, g_q$ and $g$, and respective cdfs $G_1, \ldots, G_q$ and $G$. If $h(x)$ is an increasing function, then $G_j$ is stochastically larger (smaller) than $G$ when $\beta_j > 0$ ($\beta_j < 0$).

This follows from Theorem 1 in Patil, Rao and Ratnaparkhi (1986). This result points to the flexibility of the SWS of distributions (2.1) to model different data patterns: It allows each distorted distribution to be either stochastically larger or smaller than the reference distribution, depending on the sign of the corresponding parameter $\beta_j$. In particular for each $j$, $E_{g_j}(X)$ larger and smaller than $E_g(X)$ are both possible under (2.1).

**Property 3.** Consider the SWS of distributions (2.1) with pdfs $g_1, \ldots, g_q$ and $g$, and respective cdfs $G_1, \ldots, G_q$ and $G$. If $h(x)$ is an increasing function, then for any $j, k$, $G_j$ is stochastically larger (smaller) than $G_k$ when $\beta_j > \beta_k$ ($\beta_j < \beta_k$).

This follows from Property 2 since $g_j$ is the weighted distribution obtained from $g_k$ and the weight function $\exp\{(\beta_j - \beta_k)h(x)\}$.

## 2.2 Inference for Simple Weighted Systems

### 2.2.1 Empirical Distribution

We shall often make reference to the empirical distribution obtained from a single random sample $X_1, \ldots, X_n$ from a cumulative distribution function (cdf) $G^*$. An estimator of $G^*$ is the empirical cumulative distribution, or simply the *empirical distribution*, defined for any real $x$ by

$$\hat{G}^*(x) = \frac{1}{n}\sum_{i=1}^{n} I(X_i \leq x) \tag{2.2}$$

where $I(B)$ is equal to 1 if $B$ occurs and is equal to 0 otherwise. Since the $I(X_i \leq x)$ are independent Bernoulli($G^*(x)$) random variables, their sum $n\hat{G}^*(x)$ has bin($n, G^*(x)$) distribution. It follows that $\hat{G}^*(x)$ is an unbiased estimator of $G^*(x)$, and $\text{Var}(\hat{G}^*(x)) = G^*(x)(1 - G^*(x))/n$. In addition, from the Glivenko-Cantelli theorem, $\sup\{|\hat{G}^*(x) - G^*(x)| : x \in \mathbb{R}\}$ converges to zero almost surely. The question is whether $\hat{G}^*(x)$ is optimal in some sense.

### 2.2.2 Empirical Likelihood

Here we introduce the notion of likelihood in situations where we want to make minimal assumptions about the family of distributions that generated the data. Let $X_1, \ldots, X_n$ be a random sample from an unknown distribution $G^*$. Suppose we observe the values $X_1 = x_1, \ldots, X_n = x_n$ (assumed distinct for simplicity), and let $G$ be an arbitrary cdf. The *nonparametric likelihood* of $G$ based on the above data is defined as

$$L(G) = \prod_{i=1}^{n} \left( G(x_i) - G(x_i^-) \right), \tag{2.3}$$

where $G(x_i^-) = \lim_{x \uparrow x_i} G(x)$, and $G(x_i) - G(x_i^-) \geq 0$ is the jump in $G$ at the datum $x_i$. Like in the case of parametric likelihood, $L(G)$ is the probability of the observed data under the cdf $G$, and the aim is to estimate $G^*$ by maximizing $L(G)$. It can be shown that $L(G)$ is maximized at the empirical cdf $\hat{G}$, so $\hat{G}$ is an optimal estimator in the sense of being the nonparametric maximum likelihood estimator of $G^*$. This result can be extended to the case when the $x$'s are not distinct (Kiefer and Wolfowitz 1956, Owen 2001).

We now point out a fact that greatly simplifies the process of maximizing $L(G)$, not only in the above setting, but also in more general settings to be studied later. Note that $L(G) = 0$ for any $G$ that is continuous at any of the observed values, so $L(G) > 0$ if and only if $G$ has jumps at all of the observed values. By the Lebesgue decomposition theorem, any such $G$ can be written as[1]

$$G(x) = \alpha G_1(x) + (1 - \alpha) G_0(x),$$

with $\alpha \in (0, 1]$, $G_0$ an absolutely continuous cdf and

$$G_1(x) = \sum_{i=1}^{n} p_i I(x_i \leq x),$$

where $p_i > 0$ and $\sum_{i=1}^{n} p_i = 1$. For any such $G$ we have

$$L(G) = \prod_{i=1}^{n} \alpha \left( G_1(x_i) - G_1(x_i^-) \right) \leq L(G_1),$$

with equality occuring if and only if $G = G_1$. This implies that the problem of maximizing $L(G)$ over the space of all cdfs (an infinite dimensional space)

[1] In the more general case there is a third component that involves a singular cdf. This component is quite pathological and can be ignored for all practical purposes.

is equivalent to maximizing it over the finite-dimensional space of step cdfs with jumps at the observed data

$$\mathcal{G} = \Big\{G(x) = \sum_{i=1}^{n} p_i I(x_i \leq x) : p_i > 0 \text{ and } \sum_{i=1}^{n} p_i = 1\Big\};$$

the latter is accomplished using Calculus and Lagrange multipliers.

The preceding discussion can be extended to the case when constrains are imposed on the family of distributions that generated the data, giving rise to the notion of *empirical likelihood.* As an illustration, suppose we wish to estimate $G^*$ subject to the constraint $E_G(X) = a$, with $a$ given, so we want to maximize (2.3) subject to this constraint. By the same argument as above, this is achieved by maximizing $L(G)$ in (2.3) over the finite-dimensional space

$$\mathcal{G} = \Big\{G(x) = \sum_{i=1}^{n} p_i I(x_i \leq x) : p_i > 0,\ \sum_{i=1}^{n} p_i = 1,\ \sum_{i=1}^{n} p_i x_i = a\Big\}. \tag{2.4}$$

Define $p_i \equiv G(x_i) - G(x_i^-)$, so $L(G) = \prod_{i=1}^{n} p_i$. Using Lagrange multipliers, we see that maximizing $L(G)$ over $\mathcal{G}$ in (2.4) is achieved by maximizing the function

$$H(\boldsymbol{p}, \lambda_0, \lambda_1) = \sum_{i=1}^{n} \log(p_i) - \lambda_0\Big(\sum_{i=1}^{n} p_i - 1\Big) - n\lambda_1 \sum_{i=1}^{n} p_i(x_i - a),$$

over $\boldsymbol{p} = (p_1, \ldots, p_n)' \in \mathbb{R}^n$ and $\lambda_0, \lambda_1 \in \mathbb{R}$, the latter being Lagrange multipliers.[2] Setting the partial derivatives equal to zero, the maximizer is obtained by solving the system of equations

$$\frac{\partial H}{\partial p_k} = \frac{1}{p_k} - \lambda_0 - n\lambda_1(x_k - a) = 0, \qquad k = 1, \ldots, n \tag{2.5}$$

$$\frac{\partial H}{\partial \lambda_0} = 1 - \sum_{i=1}^{n} p_i = 0 \tag{2.6}$$

$$\frac{\partial H}{\partial \lambda_1} = -n \sum_{i=1}^{n} p_i(x_i - a) = 0. \tag{2.7}$$

By additing the $n$ equations in (2.5) and using (2.6) and (2.7), we find that $\hat{\lambda}_0 = n$. Using this in (2.5), we find that

$$\hat{p}_k = \frac{1}{n} \frac{1}{1 + \hat{\lambda}_1(x_k - a)}, \qquad k = 1, \ldots, n, \tag{2.8}$$

[2]The factor $n$ accompanying the Lagrange multiplier $\lambda_1$ has no significant effect on the maximization.

where $\hat{\lambda}_1$ is obtained numerically as the solution of the equation (obtained from (2.7))

$$\frac{1}{n}\sum_{i=1}^{n}\frac{x_i - a}{1+\hat{\lambda}_1(x_i - a)} = 0.$$

In this case the maximum empirical likelihood estimator of $G$ is

$$\hat{G}(x) = \sum_{i=1}^{n}\hat{p}_i I(x_i \le x),$$

with the $\hat{p}_i$ given in (2.8). Interestingly, additional constraints on the $p_i$ still lead to expressions similar to (2.8).

The quantity $L(G)$ where the $p_i$ are subject to constraints is an example of *empirical likelihood*, and we have just seen its use in getting estimators for $G$. It would be difficult to exaggerate the importance of this tool, which we shall utilize time and again in inference for weighted systems of distributions. An early use of empirical likelihood in the context of finite populations in the presence of concomitant or auxiliary information is made in Hartley and Rao (1968). This was later generalized in Chen and Qin (1993). Using empirical likelihood, Vardi (1982) derived the nonparametric maximum likelihood estimator $\hat{G}$ of a lifetime distribution $G$ from two independent samples, one from $G$ and the other from the length-biased distribution of $G$, $\tilde{G}(x) = \int_0^x u dG(u) / \int_0^\infty u dG(u)$. He also investigated the large sample behavior of $\hat{G}$ under conditions on the sample sizes, a topic we shall deal with in Chapter 4. The closely related notion of *empirical likelihood ratio* for the construction of confidence intervals for survival functions from censored data is encountered first in Thomas and Grunkemeier (1975). The theory and applications of empirical likelihood ratios are discussed in great detail in Owen (1988, 2001).

### 2.2.3 Semiparametric Inference for Simple Weighted Systems

Moving beyond simple ANOVA, the basic idea regarding inference for SWS's addressing the four points listed at the end of Chapter 1 is to use the *empirical likelihood* (see 2.9 below), a most useful tool in semiparametric inference, where the problem is to estimate both parameters as well as probability distributions. Empirical likelihood emulates the notion of nonparametric likelihood by placing probability masses at all observed points $t_i$, an idea pioneered by Vardi (1982, 1985) in connection with inference in length bias and selection bias problems. Maximizing the empirical likelihood subject to constraints leads to closed form expressions for the probability masses, which forms the basis for kernel density estimates as will be explained at a later point. Important results concerning the empirical likelihood and in particular an extension of Wilks's Theorem (dubbed Empirical Likelihood Theorem) in nonparametric settings are due to Owen (2001). This section follows mainly Qin and Lawless (1994), Qin and Zhang (1997), and Fokianos et al. (2001).

A maximum likelihood estimator of $G(x)$ can be obtained by maximizing the empirical likelihood over the class of step cdf's with jumps at *all* the observed values $t_1, ..., t_n$ from the combined data set, subject to distributional constraints. We have seen a special case of this in Section 2.2.2. Thus, since by the density ratio model (2.1) each $g_j$ and $g$ can be evaluated at every point $t_j$ in the fused sample, the idea is to estimate $G(x)$ by a step function with numerous steps, as many as $n = n_1 + \cdots + n_q + n_m$, and estimate its jumps, denoted by $p_i$, subject to constraints. This goes beyond the corresponding empirical distribution estimate of $G(x)$ which is based on a single sample with only $n_m$ jumps. Assume from now on that $h(x)$ is a continuous function.

Accordingly, if $p_i = dG(t_i)$, the distortions have the form $\exp(\alpha_j + \beta_j h(t_i))p_i$, for $i = 1, \ldots, n$, and $j = 1, \ldots, q$, and the empirical likelihood of $(\boldsymbol{\alpha}, \boldsymbol{\beta}, G)$ based on the $m$ independent random samples is given by

$$\begin{aligned} L(\boldsymbol{\alpha}, \boldsymbol{\beta}, G) &= \Big\{ \prod_{r=1}^{q} \prod_{j=1}^{n_r} \exp(\alpha_r + \beta_r h(x_{rj})) dG(x_{rj}) \Big\} \prod_{j=1}^{n_m} dG(x_{mj}) \\ &= \prod_{i=1}^{n} p_i \cdot \prod_{j=1}^{n_1} \exp(\alpha_1 + \beta_1 h(x_{1j})) \cdots \prod_{j=1}^{n_q} \exp(\alpha_q + \beta_q h(x_{qj})) \\ &= \prod_{i=1}^{n} p_i \cdot \prod_{j=1}^{n_1} w_1(x_{1j}) \cdots \prod_{j=1}^{n_q} w_q(x_{qj}) \end{aligned} \tag{2.9}$$

where $w_m(t) \equiv 1$, and

$$w_j(t) = \exp(\alpha_j + \beta_j h(t)), \ j = 1, ..., q.$$

As in Section 2.2.2, the empirical likelihood (2.9) is maximized subject to the constrains

$$\sum_{i=1}^{n} p_i = 1, \ \sum_{i=1}^{n} p_i[w_1(t_i) - 1] = 0, \ldots, \ \sum_{i=1}^{n} p_i[w_q(t_i) - 1] = 0. \tag{2.10}$$

The constraints simply express the fact that the discrete reference probability masses and their distortions sum to 1.

To maximize the empirical likelihood (2.9) subject to the constraints (2.10), we follow a profiling procedure whereby first the $p_i$ are expressed in terms of $\boldsymbol{\alpha}, \boldsymbol{\beta}$, and then the $p_i$ are substituted back into the likelihood to produce a function of $\boldsymbol{\alpha}, \boldsymbol{\beta}$ only. Interestingly, in what follows, the relative sample sizes

$$\rho_j = n_j/n_m, \quad j = 1, ..., q$$

play an important role, both in estimation and hypothesis testing.

With fixed $\boldsymbol{\alpha}, \boldsymbol{\beta}$, the empirical likelihood (2.9) is optimized by maximizing only the product term $\prod_{i=1}^n p_i$, subject to the $m$ constraints (2.10).

Using Lagrange multipliers we obtain the appealing expression (see the Appendix)

$$p_i \equiv p_i(\boldsymbol{\alpha}, \boldsymbol{\beta}) = \frac{1}{n_m} \cdot \frac{1}{1 + \rho_1 w_1(t_i) + \cdots + \rho_q w_q(t_i)}. \tag{2.11}$$

By substituting the $p_i$ in (2.11) back in (2.9), the profile log-likelihood of $\boldsymbol{\alpha}, \boldsymbol{\beta}$ is given by,

$$\begin{aligned} \ell(\boldsymbol{\alpha}, \boldsymbol{\beta}) &= -n \log n_m - \sum_{i=1}^{n} \log[1 + \rho_1 w_1(t_i) + \cdots + \rho_q w_q(t_i)] \\ &\quad + \sum_{j=1}^{n_1} [\alpha_1 + \beta_1 h(x_{1j})] + \cdots + \sum_{j=1}^{n_q} [\alpha_q + \beta_q h(x_{qj})]. \end{aligned} \tag{2.12}$$

The score equations for $j = 1, ..., q$, are therefore,

$$\begin{aligned} \frac{\partial \ell}{\partial \alpha_j} &= -\sum_{i=1}^{n} \frac{\rho_j w_j(t_i)}{1 + \rho_1 w_1(t_i) + \cdots + \rho_q w_q(t_i)} + n_j = 0 \\ \frac{\partial \ell}{\partial \beta_j} &= -\sum_{i=1}^{n} \frac{\rho_j h(t_i) w_j(t_i)}{1 + \rho_1 w_1(t_i) + \cdots + \rho_q w_q(t_i)} + \sum_{i=1}^{n_j} h(x_{ji}) = 0. \end{aligned} \tag{2.13}$$

The solution of the score equations, which are found numerically, gives the maximum likelihood estimators $\hat{\boldsymbol{\alpha}}, \hat{\boldsymbol{\beta}}$, and consequently by substitution,

$$\begin{aligned} \hat{p}_i &= \frac{1}{n_m} \cdot \frac{1}{1 + \rho_1 \exp(\hat{\alpha}_1 + \hat{\beta}_1 h(t_i)) + \cdots + \rho_q \exp(\hat{\alpha}_q + \hat{\beta}_q h(t_i))} \\ &= \frac{1}{n_m} \cdot \frac{1}{1 + \rho_1 \hat{w}_1(t_i) + \cdots + \rho_q \hat{w}_q(t_i)} \end{aligned} \tag{2.14}$$

where $\hat{w}_j(t_i) = \exp(\hat{\alpha}_j + \hat{\beta}_j h(t_i))$. Therefore, the estimated reference cdf is

$$\hat{G}(t) = \sum_{i=1}^{n} \hat{p}_i I(t_i \leq t). \tag{2.15}$$

Likewise, the estimated distortions are $\exp(\hat{\alpha}_j + \hat{\beta}_j h(t_i))\hat{p}_i$, $j = 1, ..., q$, and the corresponding estimated distorted cdf's are given by,

$$\hat{G}_j(t) = \sum_{i=1}^{n} \exp(\hat{\alpha}_j + \hat{\beta}_j h(t_i))\hat{p}_i I(t_i \leq t), \quad j = 1, ..., q. \tag{2.16}$$

In the Appendix it is shown that the estimators $\hat{\boldsymbol{\alpha}}, \hat{\boldsymbol{\beta}}$, are consistent and asymptotically normal,

$$\sqrt{n} \begin{pmatrix} \hat{\boldsymbol{\alpha}} - \boldsymbol{\alpha}_0 \\ \hat{\boldsymbol{\beta}} - \boldsymbol{\beta}_0 \end{pmatrix} \Rightarrow \mathrm{N}(\mathbf{0}, \boldsymbol{\Sigma}) \tag{2.17}$$

as $n \to \infty$. Here $\boldsymbol{\alpha}_0$ and $\boldsymbol{\beta}_0$ denote the true parameters, and $\boldsymbol{\Sigma}$ has a "sandwich" form peculiar to semiparametric problems,

$$\boldsymbol{\Sigma} = \boldsymbol{S}^{-1} \boldsymbol{\Lambda} \boldsymbol{S}^{-1} \tag{2.18}$$

where the matrices $\boldsymbol{S}$ and $\boldsymbol{\Lambda}$ are given in the Appendix.

Very general optimality properties of the semiparametric estimates are discussed rigorously in Gilbert (2000) and Schick and Wefelmeyer (2008). Let $\hat{\boldsymbol{\theta}}_n = (\hat{\alpha}_1, ..., \hat{\alpha}_q, \hat{\beta}_1, ..., \hat{\beta}_q)$. Then Gilbert (2000) has shown that $(\hat{\boldsymbol{\theta}}_n, \hat{G})$ are asymptotically normal and efficient. Similarly, Vardi (1982) in the context of length bias, then Zhang (2000a), and more recently Lu (2007) showed that $\sqrt{n}(\hat{G} - G)$ converges to a Gaussian process with mean zero and a rather complex covariance structure. See Section 2.3.

**Example 2.2.1** The case $m = 2, q = 1$ requires a slight change in notation. For $k = 0, 1, 2$, define

$$A_k = \int \frac{h^k(t)\exp(\alpha + \beta h(t))}{1 + \rho \exp(\alpha + \beta h(t))} dG(t)$$

and,

$$\mathbf{A} = \begin{pmatrix} A_0 & A_1 \\ A_1 & A_2 \end{pmatrix}$$

With $\rho \equiv \rho_1$, Qin and Zhang (1997) showed

$$\boldsymbol{\Sigma} = \mathbf{S}^{-1}\boldsymbol{\Lambda}\mathbf{S}^{-1} = \frac{1+\rho}{\rho}\left[\mathbf{A}^{-1} - \begin{pmatrix} 1+\rho & 0 \\ 0 & 0 \end{pmatrix}\right] \tag{2.19}$$

so that under some regularity conditions and regardless of $h(x)$,

$$\sqrt{n}\begin{pmatrix} \hat{\alpha} - \alpha_0 \\ \hat{\beta} - \beta_0 \end{pmatrix} \Rightarrow \mathrm{N}(\mathbf{0}, \boldsymbol{\Sigma}) \tag{2.20}$$

where $\alpha_0, \beta_0$ are the true parameters.

As an illustration, consider the case where $\mathbf{x}_2$ is uniformly distributed in $[0,1]$, so that $g(x) = 1, 0 \le x \le 1$, and $g(x) = 0$ otherwise. Assume $\rho = 1$, $\alpha = \beta = 0$, and $h(x) = x$, and observe that when $\alpha = \beta = 0$ the two populations are identical. As $n \to \infty$, the asymptotic covariance matrix $\boldsymbol{\Sigma}$ can be obtained exactly from

$$A_k = \int_0^1 \frac{t^k \exp(\alpha + \beta t)}{1 + \rho \exp(\alpha + \beta t)} dG(t) = \frac{1}{2(k+1)}, \qquad k = 0, 1, 2$$

so that

$$\mathbf{A} = \begin{pmatrix} 1/2 & 1/4 \\ 1/4 & 1/6 \end{pmatrix}$$

and

$$\boldsymbol{\Sigma} = \frac{1+\rho}{\rho}\left[\mathbf{A}^{-1} - \begin{pmatrix} 1+\rho & 0 \\ 0 & 0 \end{pmatrix}\right] = \begin{pmatrix} 12 & -24 \\ -24 & 48 \end{pmatrix}.$$

As explained in Section 3.3.1, a sensible kernel estimate for $g(x)$ based on fused data is obtained by operating on the $\hat{p}_i$ by a kernel,

$$\hat{g}(x) = \mathrm{Kernel}(\hat{p}_i). \tag{2.21}$$

This is different than the traditional single sample kernel estimate of $g(x)$ where only the reference sample $\boldsymbol{x}_m$ is used and a kernel operates on the jumps $1/n_m$. On the other hand, in (2.15) there is no need to use smoothed $\hat{p}_i$ because $G(x)$ is estimated at numerous points $t_i$ and this tends to give relatively smooth $\hat{G}(x)$ as compared with the corresponding empirical distribution estimate obtained from a single (reference) sample.

Summarizing, by following a profiling procedure we obtained semiparametric estimators (2.21) and (2.15) for $g(x)$ and $G(x)$, respectively, and score estimating equations (2.13) for the parameters $\boldsymbol{\alpha}$ and $\boldsymbol{\beta}$. The resulting estimators $\hat{\boldsymbol{\alpha}}, \hat{\boldsymbol{\beta}}$, are consistent and asymptotically normal.

### 2.2.4 Specifying the Tilt Function

The important problem of choosing $h(x)$ is intimately linked to goodness of fit and hence can be addressed reasonably well by testing for model validity. Validation of the density ratio model can be tested by measuring the discrepancy between two different estimates of the reference $G$, the pooled semiparametric estimate $\hat{G}$ obtained from the fused data $\boldsymbol{t}$, and the corresponding empirical distribution function $\tilde{G}$ obtained from the reference sample $\boldsymbol{x}_m$ only. Since $\tilde{G}$ is model free and uniformly consistent, this procedure is sensible when sufficient amounts of data are available. In the same vein Qin and Zhang (1997) suggested for model validation the goodness of fit statistic

$$\Delta_n = \sup_t \sqrt{n}\,|\hat{G}(t) - \tilde{G}(t)|, \tag{2.22}$$

where $\hat{G}(t)$ is given in (2.15) and $\tilde{G}(t)$ is the empirical distribution from the reference sample $\boldsymbol{x}_m = (x_{m1}, \ldots, x_{mn_m})$,

$$\tilde{G}(t) = \frac{1}{n_m} \sum_{i=1}^{n_m} I(x_{mi} \leq t). \tag{2.23}$$

In addition to being a measure of goodness of fit of the density ratio model (2.1), $\Delta_n$ can help in judging the appropriateness of the tilt function $h(x)$, for a completely misspecified $h(x)$ can result in unduly large values of $\Delta_n$. In practice, however, different tilt functions may still provide harmonious conclusions as we see, for example, in the radar application in Section 2.2.7 below.

To use the statistic $\Delta_n$ we need its distribution, a problem dealt with in Section 4.2 after studying the related problem of the asymptotic behavior of $\hat{G}(t)$. In the radar example the distribution of $\Delta_n$ is approximated straightforwardly from numerous samples.

Fokianos and Kaimi (2006), while discussing the effect of a misspecified density ratio model on estimation and testing, suggest a different approach for specifying the tilt function $h(x)$ by embedding it in a family of transformations $h(x, \lambda)$ which is assumed to contain the true tilt for a particular choice of $\lambda$. In that case the true tilt can be identified provided $\lambda$ is determined correctly.

**Remark concerning general tilts.** If (2.1) is replaced by the more general tilt model

$$g_j(x) = w(x, \theta_j)g(x), \; j = 1, ..., q \tag{2.24}$$

where $w$ is a known positive function, then we may run into identifiability problems unless certain restrictions are placed on $w(x, \theta)$ (Vardi 1985, Gilbert et al. 1999). The simple weighted system model with empirical likelihood function (2.9) is a particular case of the class of selection bias models studied by Gilbert et al. (1999), in which $w_m(t) \equiv 1$ (does not depend on $(\boldsymbol{\alpha}, \boldsymbol{\beta})$) and $w_j(t) > 0$ for all $j$ and $t$. From their Theorem 2 follows that $(\boldsymbol{\alpha}, \boldsymbol{\beta}, G)$ are *identifiable*, and the same holds for the general weighted system model that will be studied in Section 2.3. Also, it was shown in Fokianos (2004) that for an appropriately chosen $w$, the solution of $p_i$ approaches asymptotically that in (2.11).

### 2.2.5 Mean Estimation

The first two moments of the tilt function $h(t)$ with respect to the reference $g$ are needed for hypothesis testing in the next section. The mean of $h(t)$,

$$\int h(t) dG(t)$$

can be estimated from the combined data using the estimator $\sum_{i=1}^{n} h(t_i)\hat{p}_i$, or by taking the average of $h(x_{m1}), ..., h(x_{mn_m})$. Interestingly, the two estimates are identical. To see this, notice that from (2.14) we can get an expression for $h(t_i)[1 - n_m\hat{p}_i]$. Summing this over $i$ and invoking (2.13) for $\beta_j$, $j = 1, ..., q$, we have

$$\sum_{i=1}^{n} h(t_i)[1 - n_m\hat{p}_i] = \sum_{i=1}^{n_1} h(x_{1i}) + \cdots + \sum_{i=1}^{n_q} h(x_{qi}) = \sum_{i=1}^{n} h(t_i) - \sum_{i=1}^{n_m} h(x_{mi})$$

since $(t_1, ..., t_n) = (\boldsymbol{x}'_1, ..., \boldsymbol{x}'_q, \boldsymbol{x}'_m)$. Therefore,

$$\sum_{i=1}^{n} h(t_i)\hat{p}_i = \frac{1}{n_m} \sum_{i=1}^{n_m} h(x_{mi}). \tag{2.25}$$

It follows from (2.25) that with $h(x) = x$ averaging the $t_i$ in the fused data gives the the sample mean in the $m$th sample (reference sample),

$$\sum_{i=1}^{n} t_i\hat{p}_i = \bar{x}_m.$$

The property (2.25) holds more generally for vector valued $\boldsymbol{h}$. This however is not the case for higher order moments of $h(t)$, and the combined estimate is not the same as the corresponding estimate from the $m$th sample.

### 2.2.6 Testing Equidistribution

Testing the equidistribution hypothesis $\mathrm{H}_0 : \beta_1 = \cdots = \beta_q = 0$, or $\mathrm{H}_0 : \boldsymbol{\beta} = \mathbf{0}$, that all the $m$ distributions are equal, can proceed in several ways. For example, one possibility is to apply the score test using the score equations (2.13) for the $\beta_j$, $j = 1, ..., q$, thus eliminating the need to evaluate $\hat{\boldsymbol{\beta}}$. Accordingly, under $\mathrm{H}_0 : \boldsymbol{\beta} = \mathbf{0}$, the score equations reduce to

$$\frac{\partial \ell}{\partial \beta_j}|_{\boldsymbol{\beta}=\mathbf{0}} = n_j \{\overline{h(x_j)} - \overline{h(t)}\},\ j = 1, ..., q,$$

where $\overline{h(x_j)}$ is the $j$th sample average and $\overline{h(t)}$ is the combined sample average. The basis for the test is the fact $E[\partial \ell / \partial \boldsymbol{\beta}] = \mathbf{0}$, which implies that the score equations should themselves be close to zero as well. However, a more direct alternative relies on the asymptotic properties of $\hat{\boldsymbol{\beta}}$.

A useful test can be constructed in terms of the relative sample sizes $\rho_j$. Consider the $q \times q$ matrix $\boldsymbol{A}_{11}$ whose $j$th diagonal element is

$$\frac{\rho_j [1 + \sum_{k \neq j}^q \rho_k]}{[1 + \sum_{k=1}^q \rho_k]^2}$$

and otherwise for $j \neq j'$, element $jj'$ is

$$\frac{-\rho_j \rho_{j'}}{[1 + \sum_{k=1}^q \rho_k]^2}.$$

For $m = 2, q = 1$, $\boldsymbol{A}_{11}$ reduces to a scalar $\rho_1/(1+\rho_1)^2$. In general, the elements of $\boldsymbol{A}_{11}$ are bounded by 1 and the matrix is nonsingular,

$$|\boldsymbol{A}_{11}| = \frac{\prod_{k=1}^q \rho_k}{[1 + \sum_{k=1}^q \rho_k]^m} > 0.$$

Then, under the equidistribution hypothesis $\mathrm{H}_0 : \beta_1 = \cdots = \beta_q = 0$, we have

$$\sqrt{n}\hat{\boldsymbol{\beta}} \Rightarrow \mathrm{N}\left(\mathbf{0}, \frac{1}{\mathrm{Var}(h(T))} \boldsymbol{A}_{11}^{-1}\right) \tag{2.26}$$

where $\mathrm{Var}(h(T))$ is the variance of $h(T)$ with respect to the reference distribution,

$$\mathrm{Var}(h(T)) \ = \ \int h^2(t) dG(t) - \left(\int h(t) dG(t)\right)^2.$$

It follows that for sufficiently large $n$, under $\mathrm{H}_0$

$$\mathcal{X}_1 = n\mathrm{Var}(h(T))\hat{\boldsymbol{\beta}}'\boldsymbol{A}_{11}\hat{\boldsymbol{\beta}} \tag{2.27}$$

is approximately distributed as $\chi^2(q)$, and $\mathrm{H}_0$ can be rejected for large values of $\mathcal{X}_1$. This holds for large $n$ even when $\mathrm{Var}(h(T))$ is replaced by its estimate

$$\hat{\mathrm{V}}\mathrm{ar}(h(T)) = \sum_{i=1}^{n} h^2(t_i)\hat{p}_i - \left(\sum_{i=1}^{n} h(t_i)\hat{p}_i\right)^2 . \tag{2.28}$$

**Comparison with the $t$-Test**

A statistical test is considered more powerful than a rival test if it requires a smaller sample size to achieve the same power for a given alternative, provided both tests have the same significance level. The relative efficiency is measured in terms of the ratio of sample sizes. Thus, the power efficiency of test 1 relative to test 2 is the ratio $N_2/N_1$ where $N_1$ is the number of observations needed by test 1 to achieve the same power as test 2 with $N_2$ observations. It is interesting to compare the efficiency of the less known (properly modified) $\mathcal{X}_1$-test relative to the widely used two-sample $t$-test using normal as well as non-normal data.

The asymptotic power efficiency of the $\mathcal{X}_1$-test with $m = 2$ relative to the two-sample $t$-test has been studied rigorously in Gagnon (2005), and Gagnon et al. (2008). To be able to compare the two tests, Gagnon (2005) used the compatible (with the $t$-test) "square root" version of $\mathcal{X}_1$ for $q = 1, m = 2$, and $\boldsymbol{A}_{11} = \rho_1/(1+\rho_1)^2$, by which we mean the 'two-sided' test based on the test statistic

$$\tilde{\mathrm{Z}}_n \equiv \sqrt{n}\frac{\sqrt{\rho_1}}{(1+\rho_1)}\sqrt{\hat{\mathrm{V}}\mathrm{ar}(h(T))} \times \hat{\beta} . \tag{2.29}$$

As an example, consider the Gamma$(a, b)$ distribution with shape parameter $a$ and scale $b$. In the first case $a = 1$, $g_1 \sim$ Gamma$(1, b)$, the reference is $g \sim$ Gamma$(1, 3)$, and $h(x) = x$. Relative efficiency curves $N_t/N_z$ as functions of $b$ are shown in Figure 2.1 for several power values and significance level .05. The $\tilde{\mathrm{Z}}_n$ test dominates the $t$-test for power values $.7, .8, .9$, and $b < 3$, and for power values approaching 1 and $b > 3$.

In the second case $b = 1$, $g_1 \sim$ Gamma$(a, 1)$, the reference is $g \sim$ Gamma$(3, 1)$, and $h(x) = \log x$. From Figure 2.2 we see that in this case $\tilde{\mathrm{Z}}_n$ dominates the $t$-test most of the time.

Similar calculations show that the $t$-test always dominates the $\tilde{\mathrm{Z}}_n$-test in the normal case with $h(x) = x$, and that the situation is reversed for the

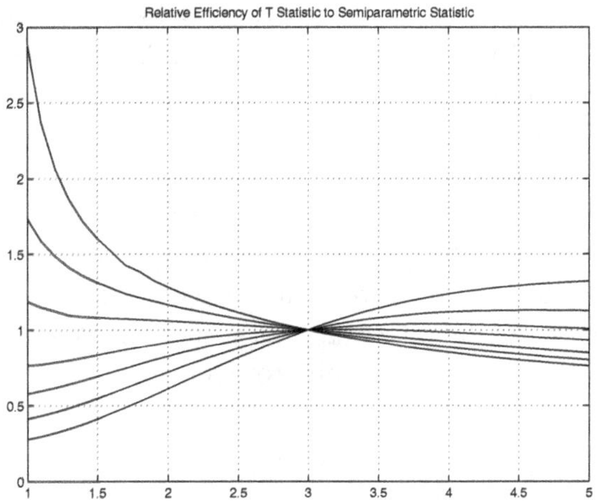


Figure 2.1: Relative efficiency curves $N_t/N_z$ versus $b$, when $g_1 \sim \text{Gamma}(1,b)$, $g \sim \text{Gamma}(1,3)$, for level .05. The curves starting from top left correspond to power values of $\gamma = .7, .8, .9, .99, .9999, .9999999999999999, 1$. $h(x) = x$. Source: Gagnon (2005).

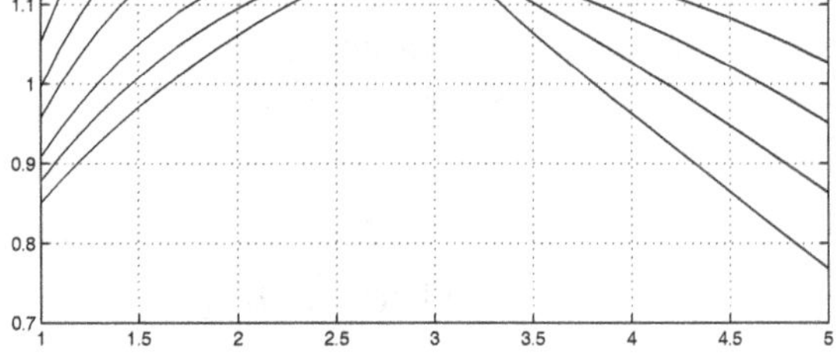

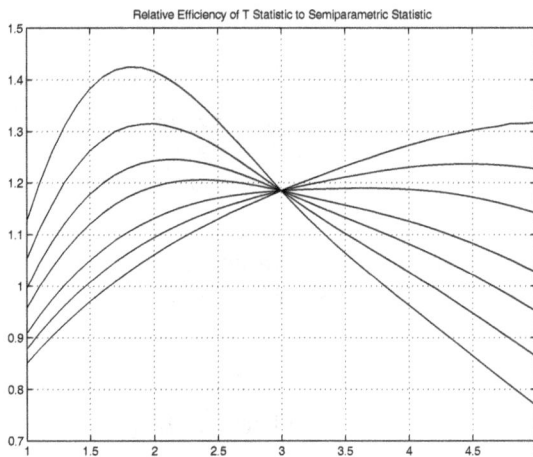

Figure 2.2: Relative efficiency curves $N_t/N_z$ versus $a$, when $g_1 \sim \text{Gamma}(a,1)$, $g \sim \text{Gamma}(3,1)$, for level .05. The curves starting from bottom left correspond to power values of $\gamma = .7, .8, .9, .99, .9999, .9999999999999999, 1$. $h(x) = \log x$. Source: Gagnon (2005).

lognormal case where the $\tilde{Z}_n$-test with $h(x) = \log x$ always dominates the $t$-test, as we also conclude from Figures 2.3 and 2.4, respectively. Thus, the power efficiency computed for several special cases points to an intriguing behavior where one test can be more efficient than the other over a certain parameter range and less efficient over some other range, or where one test completely dominates the other. However, as far as the business at hand, here is another example where the semiparametric paradigm gives rise to a useful alternative which rivals a well-entrenched statistical procedure. That is, the previous cases highlight situations where the $\tilde{Z}_n$-test, the two sided manifestation of the of $\mathcal{X}_1$-test, dominates the $t$-test.

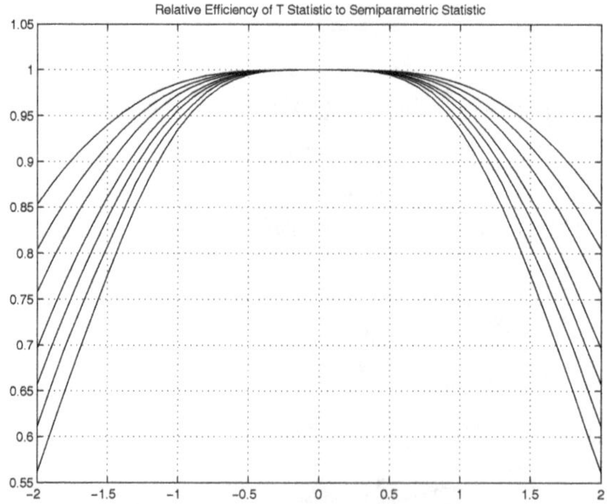


Figure 2.3: Relative efficiency curves $N_t/N_z$ versus $\mu_1 - \mu_2$, when $g_1 \sim N(\mu_1, 1)$, $g \sim N(\mu_2, 1)$, for level $\alpha = .05$. The curves, starting from the top, correspond to different power values of $\gamma = .7, .8, .9, .99, .9999, .9999999999999999, 1$. $h(x) = x$. Source: Gagnon (2005).

### 2.2.7 Application to Radar Meteorology

To illustrate the semiparametric approach described hitherto, the $\mathcal{X}_1$ test is applied in a two-sample ($m = 2$) radar problem, distinguishing between two precipitation radars. In addition to the test results we also obtain kernel density estimates by smoothing the $\hat{p}_i$, a procedure to be discussed at a later chapter. The analysis illustrated here in terms of radars is also applicable to different scenarios involving different data producing instruments or algorithms.

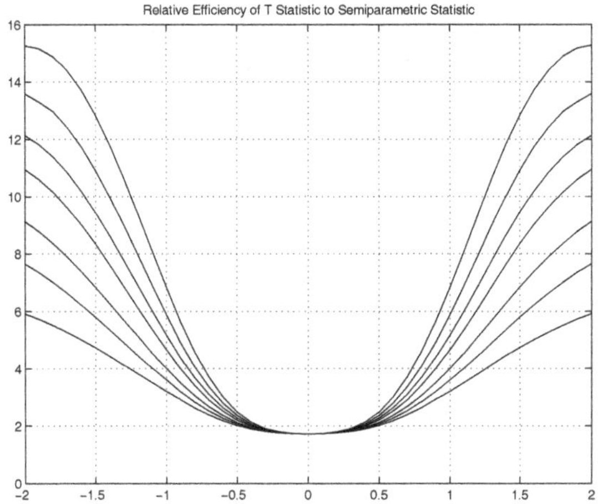


Figure 2.4: Relative Efficiency curves $N_t/N_z$ versus $\mu_{l1}$, when $g_1 \sim \text{LN}(\mu_{l1}, 1)$, $g \sim \text{LN}(0, 1)$, for level .05. The curves starting from the bottom left correspond to power values of $\gamma = .7, .8, .9, .99, .9999, .9999999999999999, 1$. $h(x) = \log x$. Source: Gagnon (2005).

Experimental radar reflectivity data were obtained from two radars deployed during NASA's Tropical Rainfall Measuring Mission (TRMM) Kwajalein Experiment (KWAJEX), held July 15 - September 12, 1999, in the Republic of the Marshall Islands. An S-band radar was located on Kwajalein Island at the southern end of the Kwajalein Atoll, and a C-band radar was on board the NOAA ship Ronald H. Brown (RHB). Different calibrations were applied to the two radars, and their spherical data were gridded to the same 1 km $\times$ 1 km $\times$ 1 km Cartesian grid. More details about the data are given in Kedem et al. (2004). We mainly deal with the first phase of the experiment, referring to the radars as "Kwajalein" and "Brown".

The data from each radar were sampled randomly to produce random samples $\boldsymbol{x}_1$ (Kwajalein sample) and $\boldsymbol{x}_2$ (Brown reference sample). The question is whether or not the two radars, or their algorithms, produce equidistributed reflectivity data by fusing $\boldsymbol{x}_1$ and $\boldsymbol{x}_2$. The analysis was repeated several times, each time with different pairs $\boldsymbol{x}_1, \boldsymbol{x}_2$, to assess the consistency of the method.

Following model (2.1) with $m = 2, q = 1, n_1 = n_2 = 500$, and $g(x)$ from Brown as the reference density, the combined data $t_1, ..., t_{1000}$ were used in the estimation of the parameters $\alpha_1, \beta_1$ and the reference density $g(x)$ using $h(x) = x$ and also $h(x) = \log x$. Radar data (or functions thereof) are often assumed distributed as lognormal or gamma, both of which motivate a logarithmic tilt function $h(x) = \log x$ for some combination of their parameters. The simplest choice of $h(x) = x$ could serve as a benchmark.

To lend support to the two choices of the tilt function, the distribution of $\Delta_n$ in (2.22) was approximated from 1000 different reflectivity samples each of size 500 obtained with replacement from large first phase records. The $p$-value for $h(x) = x$ and $\Delta_n = 1.299119$ is 0.559. For $h(x) = \log x$ and $\Delta_n = 1.762905$ the $p$-value is 0.653. Thus, based on the $p$-values both choices are sensible. However, admittedly, both choices are not optimal and are somewhat arbitrary, even if sensible and useful. In the present radar data example, both choices give very similar test results.

Table 2.1 provides typical parameter estimates from four pairs of independent samples, first with $h(x) = x$ and then with $h(x) = \log(x)$. The $\mathcal{X}_1$ test suggests, as expressed by small $p$-values, that Kwajalein very much deviates from Brown, and the hypothesis that the data are equidistributed is rejected quite conclusively. That is, the two radars operate very differently. The standard errors reported in the table were computed from $\mathbf{\Sigma}/n$, where $\mathbf{\Sigma}$ is given for $m = 2$ in Example 2.2.1, and for general $m$ in the Appendix. The standard errors match closely experimental standard errors.

Similar information is conveyed graphically in Figures 2.5 and 2.6, produced with $h(x) = x$ and $h(x) = \log x$, respectively, where we see discrepancies between $\hat{G}$ and $\hat{G}_1$ (dashed), and between kernel smoothed $\hat{g}$ and $\hat{g}_1$ (dashed). The figures also show overlays of $\hat{g}$ and $\hat{g}_1$ on the corresponding histograms where we see good fits, and hence an indication that the method "works". The fact that we see close fits between the $g$'s and the corresponding histograms is quite remarkable, for no histograms were used in the semiparametric analysis. We shall see this repeated time and again in various applications. Apparently, with either choice of $h(x)$, Figures 2.5 and 2.6 appear quite similar.

On the other hand, still with $n_1 = n_2 = 500$, from Table 2.2 we can see that when both samples come from the same Brown radar there is a dramatic decrease in the $\mathcal{X}_1$ values and consequently an appreciable increase in the corresponding $p$-values. In other words, the test recognizes quite decisively the fact that the data were generated by the same algorithm. Figure 2.7 conveys a similar message as $\hat{G}$ and $\hat{G}_1$ practically coincide. Similarly, with three (second phase) samples from Kwajalein, the hypothesis of equidistribution is accepted quite conclusively regardless of the choice of $h(x)$ as seen from the $p$-values in Table 2.3.

Table 2.1: Kwajalein, Brown (reference). Typical $\hat{\alpha}_1, \hat{\beta}_1$ values from four different $(\boldsymbol{x}_1, \boldsymbol{x}_2)$ samples, and typical $p$-values from $\mathcal{X}_1$. For $h(x) = x$ the approximate standard errors of $(\hat{\alpha}_1, \hat{\beta}_1)$ are $(0.212, 0.007)$, and for $h(x) = \log(x)$ they are $(0.493, 0.148)$. $n_1 = n_2 = 500$. The hypothesis that the data come from the same radar (algorithm) is rejected quite conclusively.

| $h(x)$ | $\hat{\alpha}_1$ | $\hat{\beta}_1$ | $\mathcal{X}_1$ | p-value |
|---|---|---|---|---|
| $x$ | 0.784 | -0.027 | 14.503 | 1.399e-03 |
| | 1.244 | -0.042 | 33.476 | 7.216e-09 |
| | 0.707 | -0.024 | 12.204 | 4.768e-04 |
| | 0.909 | -0.030 | 17.292 | 3.206e-05 |
| $\log(x)$ | 1.319 | -0.396 | 6.520 | 0.011 |
| | 1.908 | -0.575 | 12.744 | 3.572e-04 |
| | 1.871 | -0.562 | 11.202 | 8.169e-04 |
| | 2.050 | -0.621 | 16.510 | 4.838e-05 |

Table 2.2: Brown, Brown (reference). Typical $\hat{\alpha}_1, \hat{\beta}_1$ values from four different $(\boldsymbol{x}_2, \boldsymbol{x}_2)$ samples, and typical $p$-values from $\mathcal{X}_1$. For $h(x) = x$ the approximate standard errors of $(\hat{\alpha}_1, \hat{\beta}_1)$ are $(0.212, 0.007)$, and for $h(x) = \log(x)$ they are $(0.449, 0.137)$. $n_1 = n_2 = 500$. The hypothesis that the data come from the same radar (algorithm) is accepted quite conclusively.

| $h(x)$ | $\hat{\alpha}_1$ | $\hat{\beta}_1$ | $\mathcal{X}_1$ | p-value |
|---|---|---|---|---|
| $x$ | -0.078 | 0.003 | 0.140 | 0.709 |
| | 0.005 | -0.000 | 0.001 | 0.980 |
| | -0.139 | 0.005 | 0.457 | 0.499 |
| | 0.112 | -0.004 | 0.274 | 0.601 |
| $\log(x)$ | -0.584 | 0.175 | 1.723 | 0.189 |
| | 0.095 | -0.028 | 0.042 | 0.838 |
| | -0.225 | 0.067 | 0.250 | 0.617 |
| | -0.027 | 0.008 | 0.003 | 0.959 |

Table 2.3: Kwajalein, Kwajalein, Kwajalein (reference). Typical $\hat{\alpha}_1, \hat{\alpha}_2, \hat{\beta}_1, \hat{\beta}_2$ values from four different $(\boldsymbol{x}_1, \boldsymbol{x}_1, \boldsymbol{x}_1)$ samples, and typical $p$-values from $\mathcal{X}_1$. $m = 3$, $n_1 = n_2 = n_3 = 500$. The hypothesis that the data come from the same radar (algorithm) is accepted quite conclusively.

| Data | $\hat{\alpha}_1$ | $\hat{\alpha}_2$ | $\hat{\beta}_1$ | $\hat{\beta}_2$ | $\mathcal{X}_1$ | p-value |
|---|---|---|---|---|---|---|
| $h(x) = x$ | | | | | | |
| 1 | 0.108 | 0.049 | -0.003 | -0.002 | 0.283 | 0.868 |
| 2 | 0.065 | -0.003 | -0.002 | 0.000 | 0.135 | 0.935 |
| 3 | 0.227 | -0.041 | -0.007 | 0.001 | 1.896 | 0.388 |
| 4 | 0.239 | -0.220 | -0.008 | 0.007 | 4.707 | 0.095 |
| $h(x) = \log x$ | | | | | | |
| 1 | 0.453 | 2.278 | -0.132 | -0.665 | 1.929 | 0.381 |
| 2 | -0.792 | -0.223 | 0.231 | 0.065 | 0.250 | 0.882 |
| 3 | -0.359 | 0.735 | 0.105 | -0.215 | 0.553 | 0.758 |
| 4 | 1.665 | 1.246 | -0.485 | -0.363 | 1.014 | 0.602 |

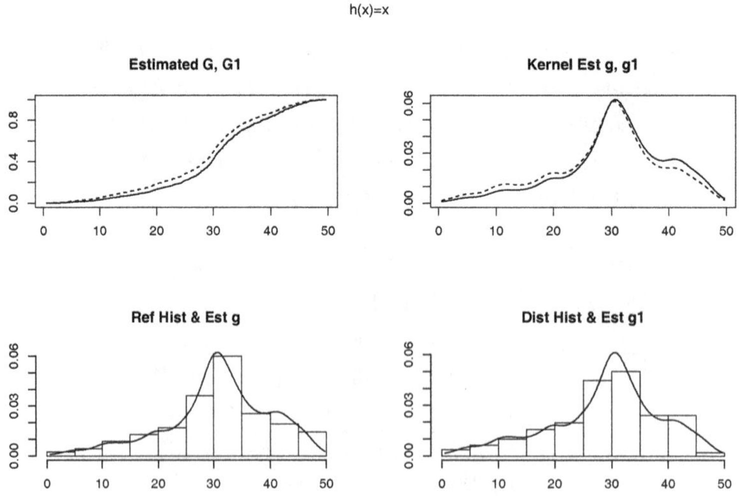


Figure 2.5: Estimated $G, G_1, g, g_1$ from Kwajalein, Brown (reference), and comparison with the corresponding histograms. $h(x) = x$, $n_1 = n_2 = 500$.

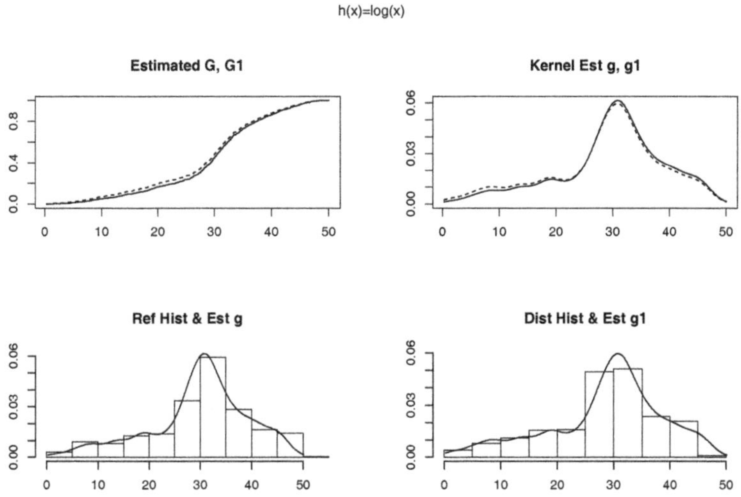


Figure 2.6: Estimated $G, G_1, g, g_1$ from Kwajalein, Brown (reference), and comparison with the corresponding histograms. $h(x) = \log x$, $n_1 = n_2 = 500$.

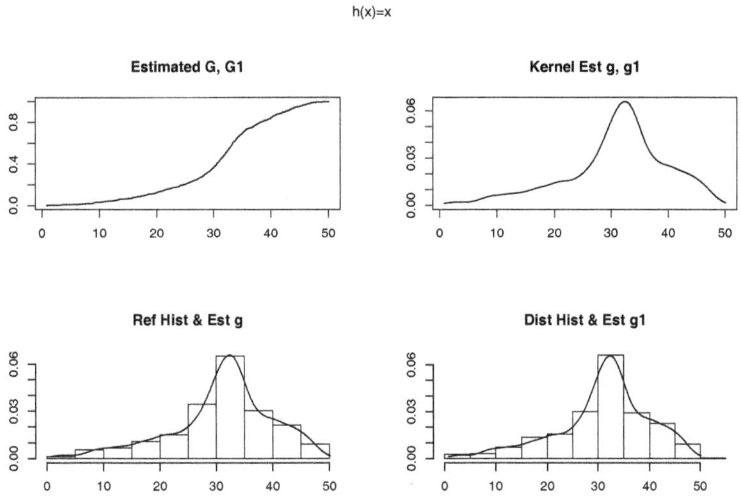


Figure 2.7: Estimated $G, G_1, g, g_1$ from Brown, Brown (reference), and comparison with the corresponding histograms. $h(x) = x$, $n_1 = n_2 = 500$.

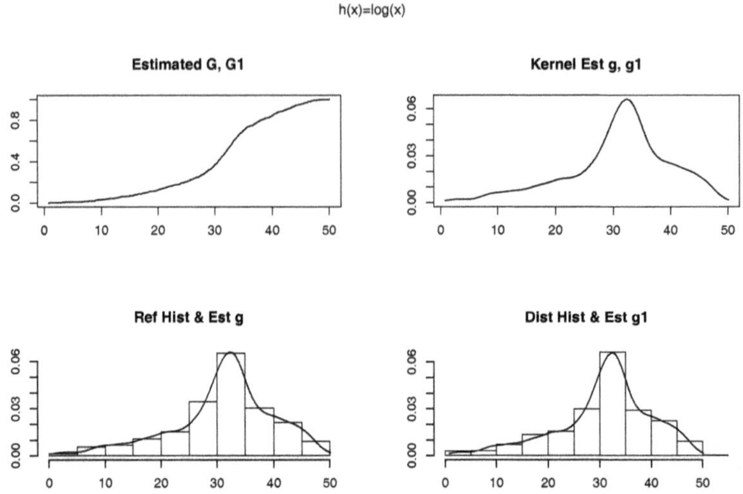


Figure 2.8: Estimated $G, G_1, g, g_1$ from Brown, Brown (reference), and comparison with the corresponding histograms. $h(x) = \log x$, $n_1 = n_2 = 500$.

The radar example is typical of a wide range of situations where the problem is to test system equilibrium such as the equilibrium of a system of undersea acoustical sensors spread over an area, all measuring the same quantities. The hypothesis of equilibrium or balance or symmetry or in short "no target" can be formulated as that of an equidistribution hypothesis, and when the hypothesis is rejected it signals system disturbance tantamount to target detection. The radar example underscores the sensitivity and usefulness of the semiparametric paradigm to such problems.

### 2.2.8 Discussion and Complements

The main point of the semiparametric paradigm discussed here is that the reference cdf $G(x)$ is estimated from many samples giving an improved estimate as compared with the empirical cdf which is obtained from the reference sample only. This fact has been addressed carefully by several authors. In particular, Gilbert (2000) has shown that the estimates $(\hat{\alpha}_1, ..., \hat{\alpha}_q, \hat{\beta}_1, ..., \hat{\beta}_q)$ and $\hat{G}$ are jointly asymptotically normal and efficient. In the same vein, Zhang (2000b) has shown that quantile estimates obtained by the semiparametric method from both case and control samples are more efficient than estimates that are based on the control sample only, ignoring the case information. More recently Fokianos (2004), Qin and Zhang (2005), and Voulgaraki et al. (2012) showed that by merging information following the semiparametric paradigm we obtain improved kernel density estimates with the same bias as the traditional kernel density estimates but with smaller asymptotic variance. Our data analysis below supports this claim. Moreover, merging information in this way can result in powerful tests for distribution equality. See Fokianos et al. (2001), Gagnon (2005), Kedem and Wen (2007), and Wen and Kedem (2009).

It can be shown that the estimates $\hat{\alpha}$'s and $\hat{\beta}$'s are asymptotically normal with a covariance structure depending on functionals of $G$. See Lu (2007), Qin and Zhang (1997), Zhang (2000b), and the appendix to this chapter. In addition, regarding the uncertainty in $\hat{G}$, Zhang (2000b) and more recently Lu (2007) showed in the multiple sample case that $\sqrt{n}(\hat{G} - G)$ converges to a Gaussian process with a complicated covariance structure.

It is important to note that the method depends on the choice of the tilt, and in particular the choice of the distortion function $h(\cdot)$. An unreasonable choice of $h(\cdot)$ is tantamount to a misspecified density ratio model, an issue dealt with in Fokianos and Kaimi (2006). It is suggested there to embed $h(\cdot)$ in a parametric family $h_\lambda(\cdot)$ where $\lambda$ is chosen optimally from likelihood considerations as is done in the Box-Cox family of transformations. The idea is sensible but could lead to the problem of the choice of the family of transformations itself. At present the problem of choosing $h(\cdot)$ is still largely open, however, as was demonstrated in the previous radar data example in Section 2.2.7, there are cases where different choices of $h(\cdot)$ still lead to the same conclusions.

Extending work by Keziou and Leoni-Aubin (2007), Cai (2014) proposed to base the inference about the density ratio model on a modification of the empirical likelihood (2.9) that is called *dual empirical likelihood.* The latter avoids the non-regularity of the empirical likelihood at $\boldsymbol{\beta} = \mathbf{0}$. In particular, Cai (2014) proposed to test the equidistribution (and other) hypotheses using a likelihood ratio test based on the dual empirical likelihood. Simulation results suggest that this test is more powerful than the (Wald) test described in Section 2.2.6.

Schick and Wefelmeyer (2008) describe the historical development of semiparametric statistics emphasizing efficient estimation in models with independent and identically distributed observations, as well as time series. They define semiparametric models in a general sense as models that are neither parametric nor nonparametric. Gill, Vardi and Wellner (1988) discussed biased sampling for completely specified weight functions. An extension of the density ratio model under various censoring schemes, using a weighted empirical likelihood in two-sample semiparametric models, has been explored in detail in Ren (2008).

## 2.3 Inference for General Weighted Systems

In the preceding development $h(x)$ was a known real-valued function. The extension to the somewhat more general case where $\boldsymbol{h}(x)$ is a $p$-dimensional known vector-valued function follows in the footsteps of a scalar $h(x)$ and is straightforward. The model is now given by (1.21), that is,

$$g_j(x) = \exp\{\ \alpha_j + \boldsymbol{\beta}_j'\boldsymbol{h}(x)\}g(x), \quad j = 1, ..., q.$$

and the problem is described in Section 1.1.1.

It is convenient to define $\boldsymbol{\alpha} = (\alpha_1, \alpha_2, ..., \alpha_q)'$ and $\boldsymbol{\beta} = (\boldsymbol{\beta}_1', \boldsymbol{\beta}_2', ..., \boldsymbol{\beta}_q')'$, and as before let $p_i = dG(t_i)$. Then the empirical likelihood becomes,

$$\begin{aligned} L(\boldsymbol{\alpha}, \boldsymbol{\beta}, G) &= \prod_{i=1}^{n} p_i \prod_{j=1}^{n_1} \exp\{\alpha_1 + \boldsymbol{\beta}_1'\boldsymbol{h}(x_{1j})\} \ldots \prod_{j=1}^{n_q} \exp\{\alpha_q + \boldsymbol{\beta}_q'\boldsymbol{h}(x_{qj})\} \\ &= \prod_{i=1}^{n} p_i \prod_{j=1}^{n_1} w_1(x_{1j}) \cdots \prod_{j=1}^{n_q} w_q(x_{qj}) \end{aligned} \tag{2.30}$$

where now $w_j(t) = \exp\{\alpha_j + \boldsymbol{\beta}_j'\boldsymbol{h}(t)\}$. Subject to the constraints (2.10), we obtain identical expressions as before, except that now the $\boldsymbol{\beta}_j$ are $p$-dimensional vectors and $\boldsymbol{h}(x)$ is vector valued. Thus, the profile log-likelihood of $\boldsymbol{\alpha}$ and $\boldsymbol{\beta}$ is given by

$$\begin{aligned} \ell(\boldsymbol{\alpha}, \boldsymbol{\beta}) = -n \log n_m &- \sum_{i=1}^{n} \log[1 + \rho_1 w_1(t_i) + \ldots + \rho_q w_q(t_i)] \\ &+ \sum_{i=1}^{q}\sum_{j=1}^{n_i} (\alpha_i + \boldsymbol{\beta}_i'\boldsymbol{h}(x_{ij})) \end{aligned} \tag{2.31}$$

and the score equations for the maximum likelihood estimators $\hat{\alpha}_j$ and $\hat{\boldsymbol{\beta}}_j$, $j = 1, ..., q$, are as in (2.13) except that now $\boldsymbol{h}(t)$ is vector valued, so that $w_j(t) = \exp(\alpha_j + \boldsymbol{\beta}'\boldsymbol{h}(t))$, and

$$\begin{aligned} \frac{\partial \ell}{\partial \alpha_j} &= -\sum_{i=1}^{n} \frac{\rho_j w_j(t_i)}{1 + \rho_1 w_1(t_i) + \cdots + \rho_q w_q(t_i)} + n_j = 0 \\ \frac{\partial \ell}{\partial \boldsymbol{\beta}_j} &= -\sum_{i=1}^{n} \frac{\rho_j w_j(t_i)\boldsymbol{h}(t_i)}{1 + \rho_1 w_1(t_i) + \cdots + \rho_q w_q(t_i)} + \sum_{i=1}^{n_j} \boldsymbol{h}(x_{ji}) = \mathbf{0}. \end{aligned} \tag{2.32}$$

As before it follows from (2.31) that the maximum likelihood estimators $\hat{\boldsymbol{\alpha}}, \hat{\boldsymbol{\beta}}$ are asymptotically normal as $n \to \infty$,

$$\sqrt{n}\begin{pmatrix} \hat{\boldsymbol{\alpha}} - \boldsymbol{\alpha}_0 \\ \hat{\boldsymbol{\beta}} - \boldsymbol{\beta}_0 \end{pmatrix} \Rightarrow \mathrm{N}(\mathbf{0}, \boldsymbol{\Sigma}) \tag{2.33}$$

where $\boldsymbol{\Sigma} = \boldsymbol{S}^{-1}\boldsymbol{\Lambda}\boldsymbol{S}^{-1}$ is given in the Appendix in Section 2.4. In particular, as explained in the Appendix,

$$\hat{p}_i = \frac{1}{n_m} \cdot \frac{1}{1 + \rho_1 \exp\{\hat{\alpha}_1 + \hat{\boldsymbol{\beta}}_1' \boldsymbol{h}(t_i)\} + \cdots + \rho_q \exp\{\hat{\alpha}_q + \hat{\boldsymbol{\beta}}_q' \boldsymbol{h}(t_i)\}} \tag{2.34}$$

$i = 1, \ldots, n$, and

$$\hat{G}(x) = \sum_{i=1}^{n} \hat{p}_i I(t_i \leq x). \tag{2.35}$$

Likewise, the estimated distortions are $\exp(\hat{\alpha}_j + \hat{\boldsymbol{\beta}}_j' \boldsymbol{h}(t_i))\hat{p}_i$, $j = 1, \ldots, q$, and the corresponding estimated cdf's are given by,

$$\hat{G}_j(t) = \sum_{i=1}^{n} \exp(\hat{\alpha}_j + \hat{\boldsymbol{\beta}}_j' \boldsymbol{h}(t_i))\hat{p}_i I(t_i \leq t), \quad j = 1, \ldots, q. \tag{2.36}$$

### 2.3.1 Hypothesis Testing

As noted before, the hypothesis $\mathrm{H}_0 : \boldsymbol{\beta} = \mathbf{0}$ implies distributional homogeneity: $g_1(x) = g_2(x) = \ldots = g_m(x) \equiv g(x)$. To test $\mathrm{H}_0$, the extension of $\mathcal{X}_1$ with the same $\boldsymbol{A}_{11}$ runs as follows. Under $\mathrm{H}_0$, we deduce from the Appendix

$$\boldsymbol{S} = \begin{pmatrix} \boldsymbol{A}_{11} & \boldsymbol{A}_{11} \otimes E[\boldsymbol{h}'(T)] \\ \boldsymbol{A}_{11} \otimes E[\boldsymbol{h}(T)] & \boldsymbol{A}_{11} \otimes E[\boldsymbol{h}(T)\boldsymbol{h}'(T)] \end{pmatrix}$$

and

$$\boldsymbol{\Lambda} = \begin{pmatrix} \mathbf{0} & \mathbf{0} \\ \mathbf{0} & \mathbf{A}_{11} \otimes Var[\boldsymbol{h}(T)] \end{pmatrix}$$

where all moments are evaluated with respect to the reference distribution, and $\otimes$ means Kronecker product. The sub-matrices defining $\boldsymbol{S}$ and $\boldsymbol{\Lambda}$ have dimensions $q \times q$, $q \times qp$, $qp \times q$ and $qp \times qp$, respectively. Then $\mathcal{X}_1$ is extended to the Wald-type statistic,

$$\mathcal{X}_1 = n\hat{\boldsymbol{\beta}}'(\boldsymbol{A}_{11} \otimes Var[\boldsymbol{h}(T)])\hat{\boldsymbol{\beta}} \tag{2.37}$$

where $Var[\boldsymbol{h}(T)]$ is the covariance matrix of $\boldsymbol{h}(T)$ with respect to the reference distribution. It follows under $\mathrm{H}_0$ that $\mathcal{X}_1$ is approximately distributed as $\chi^2$ with $qp$ degrees of freedom, and $\mathrm{H}_0$ is rejected for large values.

To test equidistribution we may also use the likelihood ratio test, with test statistic

$$LR \equiv -2[\ell(\mathbf{0},\mathbf{0}) - \ell(\hat{\boldsymbol{\alpha}},\hat{\boldsymbol{\beta}})] = -\mathbf{2}\sum_{\mathbf{i=1}}^{\mathbf{n}} \log[\mathbf{1} + \rho_1 \hat{\mathbf{w}}_\mathbf{1}(\mathbf{t_i}) + \ldots + \rho_\mathbf{q} \hat{\mathbf{w}}_\mathbf{q}(\mathbf{t_i})]$$
$$+2\sum_{i=1}^{q}\sum_{j=1}^{n_i}[\hat{\alpha}_i + \hat{\boldsymbol{\beta}}_i' \boldsymbol{h}(x_{ij})] + 2n\log\left[1+\sum_{i=1}^{q}\rho_i\right]. \quad (2.38)$$

that rejects $\mathrm{H}_0$ for large values. Under $\mathrm{H}_0$, for sufficiently large samples, $LR$ is approximately distributed as $\chi^2$ with $qp$ degrees of freedom. Several authors have raised concerns about this test since $(\boldsymbol{\alpha},\boldsymbol{\beta}) = (\mathbf{0},\mathbf{0})$ is a boundary point. However, the concerns were laid to rest by Tan (2009). Moreover, power comparisons in Kedem and Wen (2007) indicate that the likelihood ratio test is slightly more powerful than the $\mathcal{X}_1$ test.

The general linear hypothesis $\mathrm{H}_0 : \boldsymbol{H\theta} = \mathbf{c}$ can be tested by means of

$$\mathcal{X}_2 = n(\boldsymbol{H}\hat{\boldsymbol{\theta}} - \mathbf{c})'(\boldsymbol{H\Sigma H}')^{-1}(\boldsymbol{H}\hat{\boldsymbol{\theta}} - \mathbf{c}) \quad (2.39)$$

where $\boldsymbol{\theta} = (\alpha_1,\ldots,\alpha_q,\boldsymbol{\beta}_1',\ldots,\boldsymbol{\beta}_q')'$, $\boldsymbol{H}$ is $p' \times [(1+p)q)]$ predetermined matrix of rank $p'$, $p' < (1+p)q$, $\mathbf{c}$ is a $p'$-dimensional vector in $\mathbb{R}^{p'}$, and $\boldsymbol{\Sigma} = \boldsymbol{S}^{-1}\boldsymbol{\Lambda}\boldsymbol{S}^{-1}$. It follows under $\mathrm{H}_0$ that $\mathcal{X}_2$ is asymptotically distributed as $\chi^2$ with $p'$ degrees of freedom provided that the inverse of $\boldsymbol{H\Sigma H}'$ exists (Sen and Singer 1993, p. 239), and $\mathrm{H}_0$ is rejected for large values.

**Example 2.3.1** An illustration of the asymptotic behavior of $\mathcal{X}_1$ and $LR$ under the hypothesis of equidistribution for $m = 2$, $\boldsymbol{h}(x) = (x, x^2)'$, $n_1 = n_2 = 1000$, and $g, g_1$ both $\mathrm{N}(0,1)$ densities is shown in Figure 2.9 in terms of QQ-plots. The model is

$$\log\frac{g_1(x)}{g(x)} = \alpha + \beta_1 x + \beta_2 x^2 .$$

and the equidistribution hypothesis is $\mathrm{H}_0 : \beta_1 = \beta_2 = 0$. The top two panels show QQ-plots from 200 runs of the quantiles from $LR$ and $\mathcal{X}_1$, respectively, against the quantiles from chi-square with 2 degrees of freedom. The bottom-left panel compares the quantiles of $\mathcal{X}_1$ and $LR$ under $\mathrm{H}_0 : \beta_1 = \beta_2 = 0$. The bottom-right panel is a QQ-plot from $LR$ (properly modified) and chi-square with 1 degree of freedom when the hypothesis is $\mathrm{H}_0 : \beta_2 = 0$.

Similarly, Figure 2.10 shows QQ-plots from 200 runs under the same model with $\boldsymbol{h}(x) = (x, x^2)'$ and $\mathrm{H}_0 : \beta_1 = \beta_2 = 0$, when $g, g_1$ are both Uniform(0, 1) densities. Additional simulations with equidistributed $g, g_1$, for example both from Gamma(3,1), give similar results.

Apparently, Figures 2.9 and 2.10 point to a good chi-square approximation as stated.

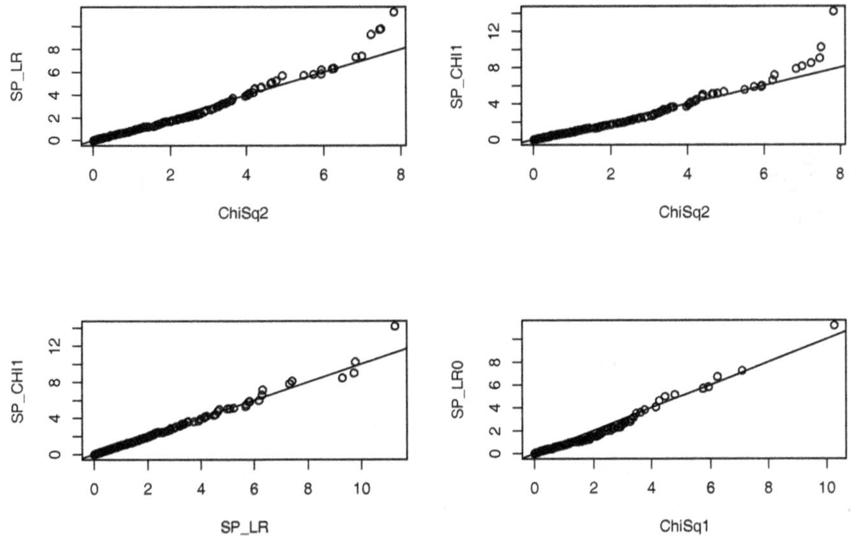


Figure 2.9: Top: QQ-plots of $LR$ and $\mathcal{X}_1$ versus $\chi_2^2$ under $H_0 : \beta_1 = \beta_2 = 0$. Bottom-left: QQ-plot of $\mathcal{X}_1$ versus $LR$ under $H_0 : \beta_1 = \beta_2 = 0$. Bottom-right: QQ-plot of $\mathcal{X}_1$ versus $\chi_1^2$ when $H_0 : \beta_2 = 0$. $g_1, g \sim \mathrm{N}(0, 1)$, $n_1 = n_2 = 1000$.

### 2.3.2 Application to Microarray Data

Our next application concerns $\log_2$ cDNA data from two lymphoma groups described in Alizadeh et al. (2000), and Qi (2002). Hierarchical clustering based on 4026 genes identified two molecularly distinct types of *diffuse large B-cell lymphoma* (DLBCL). To establish differences as well as similarities between the two groups we use replicates of expressions from different genes. There are two gene expression groups referred to as "GC" and "AC",

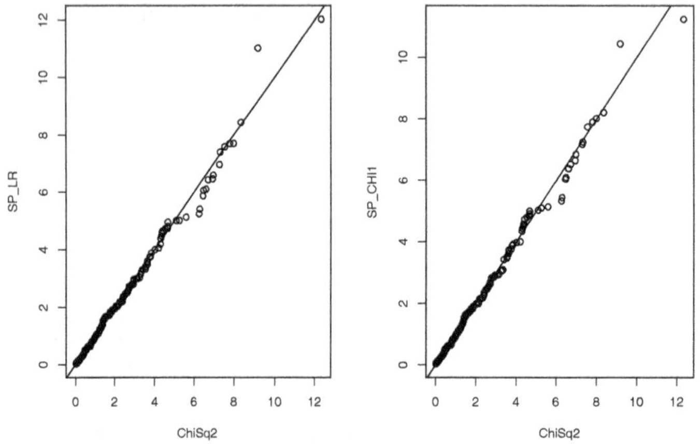


Figure 2.10: QQ-plots of $LR$ and $\mathcal{X}_1$ versus $\chi_2^2$ under $H_0 : \beta_1 = \beta_2 = 0$. $g_1, g \sim \text{Unif}(0,1)$, $n_1 = n_2 = 1000$.

where for each gene in GC (AC) there are at most 24 (23) replicated expressions. Our analysis here focuses on six genes labeled for convenience $9, 36, 69, 100, 112, 148$, out of 148 genes which were deemed useful (Qi 2002).

Both $\mathcal{X}_1$ and the likelihood ratio test are applied here in testing equidistribution in order to study the similarity in gene behavior in the AC-GC groups, designating AC as the reference group. That is, for each given gene, the reference $g$ is associated with the AC sample.

In general, experimental results show that $\log_2$ cDNA data tend to be, roughly, symmetrically distributed, as illustrated in Figure 2.11. Hence, taking a clue from the normal case, this motivates a tilt model with $\boldsymbol{h}(x) = (x, x^2)'$. Thus, in the present case of two groups, the results of the previous section are applied with the augmented model

$$g_1(x) = \exp\{\alpha_1 + \beta_1 x + \beta_2 x^2\} g(x) \tag{2.40}$$

where $g_1$ and $g$ are the distortion (GC) and the reference probability densities (AC), respectively.

We have $\boldsymbol{h}(x) = (x, x^2)'$, $m = 2$, $q = 1$. For each gene there are $n_1 \leq 24$ data points (replicates) $(x_1, \ldots, x_{n_1}) = \boldsymbol{x}_1$ in the gene group GC, and $n_2 \leq 23$ data points $(x_1, \ldots, x_{n_2}) = \boldsymbol{x}_2$ in the gene group AC. Hence, for each gene, the combined data consist of $n = n_1 + n_2 \leq 47$ observations. The combined data are used in estimating the distortion parameters $(\alpha, \beta_1, \beta_2)$,

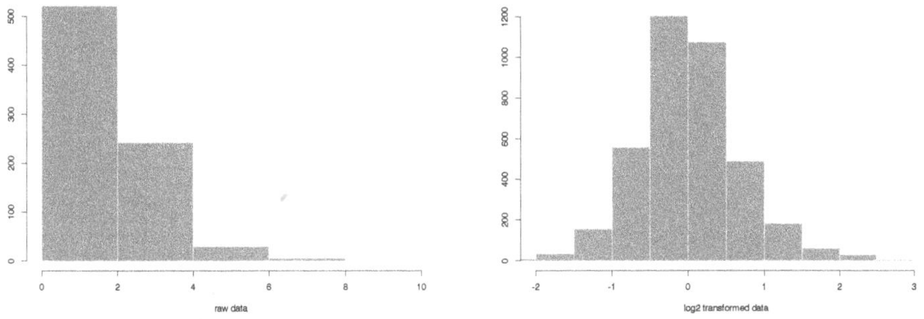


Figure 2.11: Typical histograms of microarray data based on 4026 genes. Left, raw data. Right, $\log_2$-transformed data.

the densities $g(x), g_1(x)$, and in testing the equidistribution hypothesis $H_0 : \beta_1 = \beta_2 = 0$ using both $\mathcal{X}_1$ and $LR$, and also testing $H_0 : \beta_2 = 0$ using the likelihood ratio test.

Parameter estimates and their standard errors, and $p$-values from $\mathcal{X}_1$ and the likelihood ratio test, respectively denoted by $p$-$\mathcal{X}_1$ and $p$-LR, are reported in Table 2.4. The results in the table indicate, first, that the quadratic term $x^2$ may be dropped in (2.40) and that a "linear" model with $h(x) = x$ seems appropriate. Second, in testing equidistribution both $\mathcal{X}_1$ and $LR$ give consistent results. Third, genes 9 and 100 behave somewhat similarly, in the two groups, while gene 148 behaves quite similarly. On the other hand the hypothesis of equidistribution is rejected forcefully for genes 36, 69, and 112, meaning that these genes display very different behaviors in the two groups. Gagnon et al. (2008) report similar results for $h(x) = x$.

The same information is conveyed graphically in Figure 2.12. However, the pictorial results add a useful dimension, in addition to qualitative differential information, that helps in the assessment of probabilities associated with the two groups. As an example, regarding gene 36, it is seen from Figure 2.12 that the estimated probability of an expression greater than zero is considerably higher in the tilted distribution.

A similar analysis was applied in the comparison of expression patterns of groups of genes involved in several central carbon metabolic pathways (chains of chemical reactions occurring within a cell) from *Escherichia coli* K (JM109) and *E. coli* B (BL21). These two strains respond differently to environmental factors such as glucose and oxygen concentration, and are used routinely in recombinant protein production. The analysis points to differences as well as similarities between the two E. coli strains. See Phue et al. (2007).

Table 2.4: Results of a semiparametric analysis for the indicated genes. $m = 2, q = 1, \boldsymbol{h}(x) = (x, x^2)'$.

| | | | | | | $\beta_2 = 0$ | $\beta_1 = \beta_2 = 0$ | |
|---|---|---|---|---|---|---|---|---|
| Gene | $n_1$ | $n_2$ | $\hat{\alpha}$ | $\hat{\beta}_1$ | $\hat{\beta}_2$ | $p$-LR | $p$-$\mathcal{X}_1$ | $p$-LR |
| 9 | 24 | 23 | 0.084(0.2) | 0.743(1.1) | -0.814(2.2) | 0.487 | 0.170 | 0.416 |
| 36 | 24 | 21 | 0.512(0.6) | 3.142(1.2) | -0.332(0.7) | 0.804 | 0.000 | 0.000 |
| 69 | 24 | 22 | -0.292(0.5) | 2.643(1.0) | 0.544(1.7) | 0.490 | 0.000 | 0.000 |
| 100 | 23 | 22 | -0.123(0.3) | 2.053(1.2) | -1.060(2.3) | 0.091 | 0.065 | 0.111 |
| 112 | 24 | 23 | -0.176(0.3) | 1.777(0.6) | -0.172(0.5) | 0.781 | 0.000 | 0.000 |
| 148 | 23 | 23 | -0.081(0.4) | 0.332(0.7) | 0.341(1.1) | 0.527 | 0.821 | 0.798 |

## 2.4 Appendix

### 2.4.1 Derivation of $p_i$ from the Empirical Likelihood

We follow the same notation as in Section 2.3. From (2.30),

$$L(\boldsymbol{\alpha}, \boldsymbol{\beta}, G) = \prod_{i=1}^{n} p_i \cdot \prod_{j=1}^{n_1} w_1(x_{1j}) \cdots \prod_{j=1}^{n_q} w_q(x_{qj})$$

where $w_j(t) = \exp\{\alpha_j + \boldsymbol{\beta}_j' \boldsymbol{h}(t)\}$. For fixed $\boldsymbol{\alpha} = (\alpha_1, \alpha_2, ..., \alpha_q)'$ and $\boldsymbol{\beta} = (\boldsymbol{\beta}_1', \boldsymbol{\beta}_2', ..., \boldsymbol{\beta}_q')'$, we wish to maximize the factor $\prod_{i=1}^{n} p_i$ subject to the $m$ constraints

$$\sum_{i=1}^{n} p_i = 1, \; \sum_{i=1}^{n} p_i[w_1(t_i) - 1] = 0, \ldots, \; \sum_{i=1}^{n} p_i[w_q(t_i) - 1] = 0.$$

This is equivalent to maximizing

$$\begin{aligned} \sum_{i=1}^{n} \log p_i + \lambda_0 \left[\sum_{i=1}^{n} p_i - 1\right] &- \lambda_1 \left[\sum_{i=1}^{n} p_i[w_1(t_i) - 1]\right] \\ &- \cdots - \lambda_q \left[\sum_{i=1}^{n} p_i[w_q(t_i) - 1]\right] \end{aligned}$$

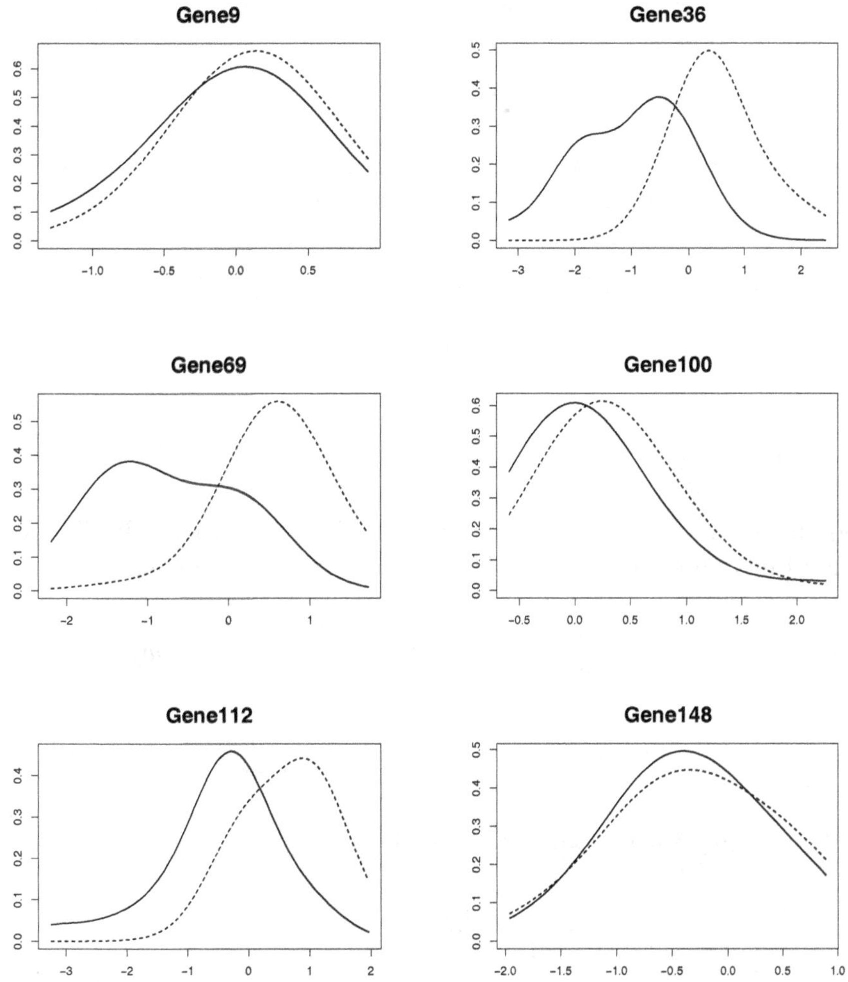


Figure 2.12: Kernel density estimates of the reference $g(x)$ (continuous line) and its distortion $g_1(x)$ (dashed line) for the indicated genes. $\boldsymbol{h}(x) = (x, x^2)'$.

where $\lambda_0, \lambda_1, \ldots, \lambda_q$ are Lagrange multipliers. Differentiating with respect to $p_i$ and equating to 0 gives,

$$1 + \lambda_0 p_i - \lambda_1 \left[p_i(w_1(t_i) - 1)\right] - \cdots - \lambda_q \left[p_i(w_q(t_i) - 1)\right] = 0,$$

so adding the previous equations for $i = 1, \ldots, n$ we obtain $\lambda_0 = -n$. Substituting this in the previous equation and solving for $p_i$ gives

$$p_i = p_i(\boldsymbol{\alpha}, \boldsymbol{\beta}) = \frac{1}{n + \lambda_1 \left[w_1(t_i) - 1\right] + \cdots + \lambda_q \left[w_q(t_i) - 1\right]}. \tag{2.41}$$

The last $q$ constraints then take the form,

$$\sum_{i=1}^{n} \frac{w_j(t_i) - 1}{n + \lambda_1 \left[w_1(t_i) - 1\right] + \cdots + \lambda_q \left[w_q(t_i) - 1\right]} = 0, \quad j = 1, \ldots, q. \tag{2.42}$$

To obtain $\lambda_1, \ldots, \lambda_q$, substitute $p_i = p_i(\boldsymbol{\alpha}, \boldsymbol{\beta})$ back into the likelihood to obtain a function of $\boldsymbol{\alpha}, \boldsymbol{\beta}$ only. The log-likelihood is then given by

$$\begin{aligned} \ell = -n \log n_m \quad &- \sum_{i=1}^{n} \log\{n + \lambda_1[w_1(t_i) - 1] + \cdots + \lambda_q[w_q(t_i) - 1]\} \\ &+ \sum_{j=1}^{n_1}[\alpha_1 + \boldsymbol{\beta}_1' \boldsymbol{h}(x_{1j})] + \cdots + \sum_{j=1}^{n_q}[\alpha_q + \boldsymbol{\beta}_q' \boldsymbol{h}(x_{qj})]. \end{aligned}$$

By differentiating $\ell$ with respect to $\alpha_1$ and equating to 0 we get the equation,

$$-\sum_{i=1}^{n} \frac{\lambda_1[w_1(t_i) - 1 + 1]}{n + \lambda_1 \left[w_1(t_i) - 1\right] + \cdots + \lambda_q \left[w_q(t_i) - 1\right]} + n_1 = 0,$$

and by invoking (2.41) and the corresponding constraint in (2.42) we get,

$$0 - \sum_{i=1}^{n} \lambda_1 p_i + n_1 = 0$$

and so $\lambda_1 = n_1$. Repeating this for $\alpha_2, \ldots, \alpha_q$ gives

$$\lambda_j = n_j, \; j = 1, \ldots, q.$$

Substitute the $\lambda_j$ in the expression for $p_i$ to obtain, after some algebra,

$$p_i = \frac{1}{n_m} \cdot \frac{1}{1 + \rho_1 w_1(t_i) + \cdots + \rho_q w_q(t_i)}$$

where $\rho_j = n_j/n_m,\ j = 1, ..., q$. Using the $\rho_j$, the log-likelihood is expressed as

$$\begin{aligned} \ell = -n \log n_m &- \sum_{i=1}^{n} \log\{1 + \rho_1 w_1(t_i) + \cdots + \rho_q w_q(t_i)\} \\ &+ \sum_{j=1}^{n_1} [\alpha_1 + \boldsymbol{\beta}_1' \boldsymbol{h}(x_{1j})] + \cdots + \sum_{j=1}^{n_q} [\alpha_q + \boldsymbol{\beta}_q' \boldsymbol{h}(x_{qj})]. \end{aligned}$$

### 2.4.2 Derivation of $\boldsymbol{S}, \boldsymbol{\Lambda}$

Assume model (1.21). The matrices $\boldsymbol{S}, \boldsymbol{\Lambda}$ are derived by repeated differentiation of (2.31). Recall that the asymptotic covariance matrix of the estimates in (2.33) is given by the product $\boldsymbol{\Sigma} = \boldsymbol{S}^{-1} \boldsymbol{\Lambda} \boldsymbol{S}^{-1}$. It is assumed throughout that all moment expressions of $\boldsymbol{h}(t)$ with respect to the reference distribution are finite.

First define

$$\nabla \equiv \left( \frac{\partial}{\partial \alpha_1}, ..., \frac{\partial}{\partial \alpha_q}, \frac{\partial}{\partial \boldsymbol{\beta}_1} ..., \frac{\partial}{\partial \boldsymbol{\beta}_q} \right)'.$$

Then $E[\nabla \ell(\alpha_1, ..., \alpha_q, \boldsymbol{\beta}_1, .., \boldsymbol{\beta}_q)] = \boldsymbol{0}$; see (4.8), (4.9). To obtain the score second moments it is convenient to define $\rho_m \equiv 1,\ w_m(t) \equiv 1$,

$$E_j[\boldsymbol{h}(T)] \equiv \int \boldsymbol{h}(t) w_j(t) dG(t)$$

and,

$$\begin{aligned} A_0(j, j') &\equiv \int \frac{w_j(t) w_{j'}(t) dG(t)}{1 + \sum_{k=1}^{q} \rho_k w_k(t)} \\ \boldsymbol{A}_1(j, j') &\equiv \int \frac{\boldsymbol{h}(t) w_j(t) w_{j'}(t) dG(t)}{1 + \sum_{k=1}^{q} \rho_k w_k(t)} \\ \boldsymbol{A}_2(j, j') &\equiv \int \frac{\boldsymbol{h}(t) \boldsymbol{h}'(t) w_j(t) w_{j'}(t) dG(t)}{1 + \sum_{k=1}^{q} \rho_k w_k(t)} \end{aligned}$$

for $j, j' = 1, ..., q$. Then, the entries in

$$\boldsymbol{\Lambda} \equiv Var\left[ \frac{1}{\sqrt{n}} \nabla \ell(\alpha_1, ..., \alpha_q, \boldsymbol{\beta}_1, ..., \boldsymbol{\beta}_q) \right]$$

are,

$$
\begin{aligned}
\frac{1}{n} Var\left(\frac{\partial \ell}{\partial \alpha_j}\right) &= \frac{\rho_j^2}{1+\sum_{k=1}^q \rho_k}\{A_0(j,j) - \sum_{r=1}^m \rho_r A_0^2(j,r)\} \\
\frac{1}{n} Cov\left(\frac{\partial \ell}{\partial \alpha_j}, \frac{\partial \ell}{\partial \alpha_{j'}}\right) &= \frac{\rho_j \rho_{j'}}{1+\sum_{k=1}^q \rho_k}\{A_0(j,j') \\
&- \sum_{r=1}^m \rho_r A_0(j,r) A_0(j',r)\} \\
\frac{1}{n} Cov\left(\frac{\partial \ell}{\partial \alpha_j}, \frac{\partial \ell}{\partial \boldsymbol{\beta}_j}\right) &= \frac{\rho_j^2}{1+\sum_{k=1}^q \rho_k}\{A_0(j,j) E_j[\boldsymbol{h}'(t)] \\
&- \sum_{r=1}^m \rho_r A_0(j,r) \boldsymbol{A}_1'(j,r)\} \\
\frac{1}{n} Cov\left(\frac{\partial \ell}{\partial \alpha_j}, \frac{\partial \ell}{\partial \boldsymbol{\beta}_{j'}}\right) &= \frac{\rho_j \rho_{j'}}{1+\sum_{k=1}^q \rho_k}\{A_0(j,j') E_{j'}[\boldsymbol{h}'(t)] \\
&- \sum_{r=1}^m \rho_r A_0(j,r) \boldsymbol{A}_1'(j',r)\} \\
\frac{1}{n} Cov\left(\frac{\partial \ell}{\partial \boldsymbol{\beta}_j}, \frac{\partial \ell}{\partial \boldsymbol{\beta}_{j'}}\right) &= \frac{\rho_j \rho_{j'}}{1+\sum_{k=1}^q \rho_k}\{-\boldsymbol{A}_2(j,j') + E_j[\boldsymbol{h}(t)] \boldsymbol{A}_1'(j,j') \\
&+ \boldsymbol{A}_1(j,j') E_{j'}[\boldsymbol{h}'(t)] \\
&- \sum_{r=1}^m \rho_r \boldsymbol{A}_1(j,r) \boldsymbol{A}_1'(j',r)\} \\
&+ \frac{1}{n}\sum_{i=1}^{n_j}\sum_{k=1}^{n_{j'}} Cov[\boldsymbol{h}(x_{ji}), \boldsymbol{h}(x_{j'k})]
\end{aligned}
$$

The last term is 0 for $j \neq j'$ and $(n_j/n)Var[\boldsymbol{h}(x_{j1})]$ for $j = j'$.

Next, as $n \to \infty$,

$$
-\frac{1}{n}\nabla\nabla' l(\alpha_1, ..., \alpha_q, \boldsymbol{\beta}_1, ..., \boldsymbol{\beta}_q) \to \boldsymbol{S}
$$

where $\boldsymbol{S}$ is a $q(1+p) \times q(1+p)$ matrix with entries corresponding to $j, j' =$

$1, ..., q,$

$$-\frac{1}{n}\frac{\partial^2 \ell}{\partial \alpha_j^2} \rightarrow \frac{\rho_j}{1+\sum_{k=1}^q \rho_k}\int \frac{[1+\sum_{k\neq j}^q \rho_k w_k(t)]w_j(t)}{1+\sum_{k=1}^q \rho_k w_k(t)}dG(t)$$

$$-\frac{1}{n}\frac{\partial^2 \ell}{\partial \alpha_j \alpha_{j'}} \rightarrow \frac{-\rho_j\rho_{j'}}{1+\sum_{k=1}^q \rho_k}\int \frac{w_j(t)w_{j'}(t)}{1+\sum_{k=1}^q \rho_k w_k(t)}dG(t)$$

$$-\frac{1}{n}\frac{\partial^2 \ell}{\partial \alpha_j \partial \boldsymbol{\beta}_j'} \rightarrow \frac{\rho_j}{1+\sum_{k=1}^q \rho_k}\int \frac{[1+\sum_{k\neq j}^q \rho_k w_k(t)]w_j(t)\boldsymbol{h}'(t)}{1+\sum_{k=1}^q \rho_k w_k(t)}dG(t)$$

$$-\frac{1}{n}\frac{\partial^2 \ell}{\partial \alpha_j \partial \boldsymbol{\beta}_{j'}'} \rightarrow \frac{-\rho_j\rho_{j'}}{1+\sum_{k=1}^q \rho_k}\int \frac{w_j(t)w_{j'}(t)\boldsymbol{h}'(t)}{1+\sum_{k=1}^q \rho_k w_k(t)}dG(t)$$

$$-\frac{1}{n}\frac{\partial^2 \ell}{\partial \boldsymbol{\beta}_j \partial \boldsymbol{\beta}_j'} \rightarrow \frac{\rho_j}{1+\sum_{k=1}^q \rho_k}\int \frac{[1+\sum_{k\neq j}^q \rho_k w_k(t)]w_j(t)\boldsymbol{h}(t)\boldsymbol{h}'(t)}{1+\sum_{k=1}^q \rho_k w_k(t)}dG(t)$$

$$-\frac{1}{n}\frac{\partial^2 \ell}{\partial \boldsymbol{\beta}_j \partial \boldsymbol{\beta}_{j'}'} \rightarrow \frac{-\rho_j\rho_{j'}}{1+\sum_{k=1}^q \rho_k}\int \frac{w_j(t)w_{j'}(t)\boldsymbol{h}(t)\boldsymbol{h}'(t)}{1+\sum_{k=1}^q \rho_k w_k(t)}dG(t)$$

In deriving the elements of $\boldsymbol{S}$ we have used the fact that averages from the fused sample $\boldsymbol{t}$ can be expressed in terms of averages from the $\boldsymbol{x}_i$ samples as follows. Assume that all $n_i \rightarrow \infty$ such that the $\rho_i = n_i/n_m$ remain fixed. Then, by summing over each sample, an application of the law of large numbers gives for function $\phi(t)$,

$$\frac{1}{n}\sum_{i=1}^{n}\phi(t_i) = \frac{n_1}{n}\frac{1}{n_1}\sum_{j=1}^{n_1}\phi(x_{1j}) + \cdots + \frac{n_q}{n}\frac{1}{n_q}\sum_{j=1}^{n_q}\phi(x_{qj}) + \frac{n_m}{n}\frac{1}{n_m}\sum_{j=1}^{n_m}\phi(x_{mj})$$

$$\rightarrow \frac{\rho_1}{1+\rho_1+\cdots+\rho_q}\int \phi(t)w_1(t)dG(t) + \cdots + \frac{1}{1+\rho_1+\cdots+\rho_q}\int \phi(t)dG(t)$$

$$= \frac{1}{1+\rho_1+\cdots+\rho_q}\left\{\int \phi(t)dG(t)[1+\rho_1 w_1(t)+\cdots+\rho_q w_q(t)]\right\}. \qquad (2.43)$$

### 2.4.3 R Code for $m = 2$, $h(x) = (x, x^2)$

In relation to the density ratio model

$$\log\frac{g_1(x)}{g(x)} = \alpha + \beta x + \gamma x^2,$$

with $m = 2$, the R function `SP2XXSQK` takes as input two samples $\boldsymbol{x}_1, \boldsymbol{x}_2$, plotting parameter `Increment`, bandwidth parameter `BandWidth` for kernel density estimates as in (2.21), and threshold $T$ for computing the upper tail probabilities $1 - G_1(T), 1 - G(T)$. Its output includes estimates $\hat{\alpha}, \hat{\beta}, \hat{\gamma}$, LR test for the hypothesis $H_0 : \gamma = 0$, $\mathcal{X}_1$ and LR tests for the hypothesis $H_0 : \beta = \gamma = 0$, a check on the value of $\hat{\alpha}$ using the relationship

$$\alpha = -\log\left\{\int \exp(\beta x + \gamma x^2) g(x) dx\right\},$$

a verification that the $p_i$ and their tilt sum to 1, the tail probabilities, $1 - G_1(T), 1 - G(T)$, and plots similar to those in Figures 2.5–2.8. The plots convey graphically how similar or dissimilar the reference and the tilted distributions are, and in this sense they complement the equidistribution hypothesis. The code can be easily modified to accommodate other tilt functions such as $h(x) = (x, \log x)$.

```
SP2XXSQK <- function(x1,x2,Increment,BandWidth,T){
### m=2,h(x)=(x,x^2). K means Kernel Density Estimate.
### Increment controls the grid at which est. g,g1,G,G1
### are evaluated for plotting purposes.
### BandWidth controls smoothness of kernel est. of g,g1.
### Reasonable values are Increment=0.05, BandWidth=0.3,0.5,1.
### T is an upper probability point: We get 1-G1(T), 1-G(T).
### G,g reference; G1,g1 tilted distributions.

###The data: m=2, x2 is the reference sample.
n1 <- length(x1); n2 <- length(x2); rho <- n1/n2
n <- n1+n2
level <- 0.05
t <- c(x1,x2) #Data fusion.

###MINUS log-likelihood
minusloglike <- function(theta) {
sum(log(1+rho*exp(theta[1] + theta[2]*(t) + theta[3]*(t)^2))) -
sum(theta[1]+theta[2]*(x1)+theta[3]*(x1)^2)}
###Maximizing loglikelihood by minimizing MINUS loglikelihood
min.func <- nlminb( start=c(-0.02,.2,.2),obj = minusloglike)

###Parameter estimates
alpha <- min.func$par[1]
beta <- min.func$par[2]
gamma <- min.func$par[3]

###Reference dist. p=dG and its distortion p1=w*dG
p <- 1/(n2*(1+rho*exp(alpha + beta*(t) + gamma*(t)^2)))
```

```
p1 <- exp(alpha + beta*(t) + gamma*(t)^2)*p

###Plots of est. G,G1,g,g1

par(mfrow=c(2,2), oma=c(0,0,4,0),cex=1.05)

###Estimated ref. cdf G(x):
G <- function(x){sum(p[t <= x])}
U <- 1-G(T)

###To Plot G over the range (min(t),max(t)):
#Modify Increment as needed. 0.05 reasonable
x <- seq(min(t),max(t),Increment)

#Creating a vector out of G for plotting.
cdf <- length(x)
for(i in 1:length(x)){
cdf[i] <- G(x[i])}

###The distorted cdf of x1:
G1 <- function(x){sum(p1[t <= x])}
U1 <- 1-G1(T)

cdf1 <- length(x)
for(i in 1:length(x)){
cdf1[i] <- G1(x[i])}

plot(x,cdf, xlab=" ",ylab=" ",type="l",lwd=1.25)
lines(x,cdf1, type="l", lty=2,lwd=1.25)
title("Estimated G, G1")

###Kernel Density Estimates of g(x) and g1(x).
ghat <- function(x){
K <- function(x){(1/sqrt(2*pi))*exp(-x^2/2)}
sum(p*K((x-t)/BandWidth)/BandWidth)}

g1hat <- function(x){
K <- function(x){(1/sqrt(2*pi))*exp(-x^2/2)}
sum(p1*K(( x-t)/BandWidth)/BandWidth)}

###Plots of ghat and g1hat
#For plots: Creating vectors from ghat and Distortted g1hat
x <- seq(min(t),max(t),Increment)
g <- rep(0,length(x))
g1 <- rep(0,length(x))
```

```
for(i in 1:length(x)){
g[i] <- ghat(x[i])
g1[i] <- g1hat(x[i])} ### Estimated pdfs.

#plots
maxgg1 <- max(c(g,g1))
plot(x,g, type="l",ylim=c(0,maxgg1),xlab=" ",ylab=" ",lwd=1.25)
lines(x,g1, type="l",lty=2,lwd=1.25)
title("Kernel Est g, g1")

###Hist of Reference Data vs Estimated Reference pdf
maxghist <- max(hist(x2,plot=FALSE)$density,g)
hist(x2,prob=T, xlab=" ",ylab=" ",lwd=1.25,
ylim=c(0,maxghist),main="Ref Hist & Est g")
lines(x,g, type="l",lwd=1.25)

###Hist of Distorted Data vs Estimated g1 pdf
maxg1hist <- max(hist(x1,plot=FALSE)$density,g1)
hist(x1,prob=T, xlab=" ",ylab=" ",lwd=1.25,
ylim=c(0,maxg1hist),main="Dist Hist & Est g1")
lines(x,g1, type="l",lwd=1.25)

#mtext("SP2XXSQ, m=2, h(x)=(x,x^2)", cex=1.2, line=1,side=3,outer=TRUE)
mtext("h(x)=(x,x^2)", cex=1.4, line=1,side=3,outer=TRUE)

###The LR Test gamma=0
logL <- -sum(log(1+rho*exp(alpha + beta*(t) + gamma*(t)^2))) +
         sum(alpha + beta*(x1) + gamma*(x1)^2)

#MINUS log-likelihood under H_0: gamma=0
minusloglike0 <- function(theta) {
sum(log(1+rho*exp(theta[1] + theta[2]*(t) + 0*(t)^2))) -
sum(theta[1]+theta[2]*(x1)+0*(x1)^2)}

#Maximizing loglikelihood under H_0 by minimizing minusloglike0
min.func0 <- nlminb( start=c(0,0),obj = minusloglike0)

#Parameter estimates under H_0: gamma=0
alpha0 <- min.func0$par[1]
beta0 <- min.func0$par[2]

#Max log-likelihood under H_0: gamma=0
logL0 <- -sum(log(1+rho*exp(alpha0 + beta0*(t) + 0*(t)^2))) +
         sum(alpha0 + beta0*(x1) + 0*(x1)^2)
```

```
#Liklihood Ratio Test Statistic for H_0: gamma=0
LR0 <- -2*(logL0 - logL)

###The Chi1 Test of Equidistribution (beta,gamma)=(0,0):
tm1 <- sum(t*p); tm2 <- sum(p*(t)^2); tm3 <- sum(p*(t)^3);
tm4 <- sum(p*(t)^4)
vart <- tm2-(tm1)^2
covtt2 <- tm3-tm1*tm2
vart2 <- tm4-(tm2)^2
VARh <- matrix(c(vart,covtt2,covtt2,vart2),ncol=2)
A11 <- rho/((1+rho)^2)
BETA <- c(beta,gamma)
chi1 <- n*A11*BETA%*%VARh%*%BETA

###The LR Test of Equidistribution (beta,gamma)=(0,0)
logLR00 <- -n*log(1+rho)
LR <- -2*(logLR00 - logL)

###Check Sum p, Sum p1
Sum_p <- sum(p); Sum_p1 <- sum(p1)

###Check: If integral relation holds
ALPHA <- -log(sum(exp(beta*t + gamma*t^2)*p))

list(alpha=alpha,beta=beta,gamma=gamma,
pval_LRgamma = 1-pchisq(LR0,1),
pval_chi1=1-pchisq(chi1,2), pval_LR=1-pchisq(LR,2),
Sum_p=Sum_p, Sum_p1=Sum_p1, ALPHA=ALPHA,
Upper_Threshold=T,Upper_G1=U1,Upper_G=U)}
```

Example:

```
> x1 <- rnorm(120,1.1,1.5)
> x2 <- rnorm(100,1.0,1.0)
> SP2XXSQK(x1,x2,0.05,0.5,2.644854)

$alpha            $beta            $gamma
[1] 0.09764964    [1] -0.791805    [1] 0.2980058
$pval_LRgamma      $pval_chi1            $pval_LR
[1] 3.131146e-05   [1,] 0.007961744      [1] 6.370608e-05
$Sum_p  $Sum_p1          $ALPHA
[1] 1   [1] 0.9999997   [1] 0.0976499
$Upper_Threshold   $Upper_G1          $Upper_G
[1] 2.644854       [1] 0.1076313      [1] 0.05084245
```

# Chapter 3

# Multivariate Extension

*"Nothing is enough for the man to whom enough is too little."*
(Epicurus, 341-270 B.C.)

## 3.1 Introduction

The previous setup can be easily generalized to multivariate data, by repeated tilting of a baseline multivariate probability density. The resulting density ratio model is indistinguishable from its univarite counterpart, except that the data are now multivariate. An important fact is that the jumps $p_i$ have the same form as those obtained earlier in the univariate case, the implication being is that we get a class of useful non-normal multivariate distributions.

The multivariate density ratio method is useful for several reasons. First, the method provides a way for determining and quantifying the differences between two or more multivariate distributions based on the joint behavior of many variables. Second, the method provides estimates of joint probabilities in case and control groups. Third, the method can be used to predict case-control results in generalized logistic regression (1.5). And fourth, we can obtain multivarite kernel density estimates which dominate the traditional kernel density estimates because more data are involved due to fusion, for example, fusion of both case and control multivariate data. As an important byproduct, the multivariate densities can be used in the estimation of conditional expectations treating one variable as a response and the rest as covariates. This is yet another tool in nonlinear regression, joining the like of Nadaraya-Watson kernel regression and other similar methods. Lastly, the implementation of the method is surprisingly simple and relatively fast at least in bivariate problems.

## 3.2 The Bivariate Case

As the general multivariate case is entirely analogous to the somewhat simpler bivariate case, it is convenient to focus first on the latter to introduce the basic construction. A clue for generalizing (1.21) is obtained from the ratio of two bivariate normal densities with different means but the same covariance matrices used in classification of multivriate vectors into one of several categoties.

Corresponding to (1.21), suppose we have $m = q + 1$ two-dimensional data sets,

$$(x_{j1}, y_{j1}), (x_{j2}, y_{j2}), \quad \ldots \quad , (x_{jn_j}, y_{jn_j}) \sim g_j(x, y), \quad j = 1, ..., q, m$$

where $g_j(x, y)$ is the probability density of $\mathrm{N}(\boldsymbol{\mu}_j, \boldsymbol{\Sigma})$, with

$$\boldsymbol{\mu}_j = \begin{pmatrix} \mu_{jx} \\ \mu_{jy} \end{pmatrix}, \quad \boldsymbol{\Sigma} = \begin{pmatrix} \sigma_{xx} & \sigma_{xy} \\ \sigma_{xy} & \sigma_{yy} \end{pmatrix}, \quad j = 1, ...m.$$

Then, choosing $g_m(x, y)$ as a reference density we have

$$\frac{g_j(x, y)}{g_m(x, y)} = \exp[(\boldsymbol{\mu}_j - \boldsymbol{\mu}_m)'\boldsymbol{\Sigma}^{-1}\boldsymbol{x} - \frac{1}{2}(\boldsymbol{\mu}_j'\boldsymbol{\Sigma}^{-1}\boldsymbol{\mu}_j - \boldsymbol{\mu}_m'\boldsymbol{\Sigma}^{-1}\boldsymbol{\mu}_m)], \tag{3.1}$$

where $\boldsymbol{x} = (x, y)'$. We see that (3.1) is a special case of the general form

$$\frac{g_j(x, y)}{g_m(x, y)} = \frac{g_j(\boldsymbol{x})}{g_m(\boldsymbol{x})} = \exp(\alpha_j + \boldsymbol{\beta}_j'\boldsymbol{x}) \tag{3.2}$$

where

$$\alpha_j = -\tfrac{1}{2}(\boldsymbol{\mu}_j'\boldsymbol{\Sigma}^{-1}\boldsymbol{\mu}_j - \boldsymbol{\mu}_m'\boldsymbol{\Sigma}^{-1}\boldsymbol{\mu}_m)$$

$$\boldsymbol{\beta}_j = \begin{pmatrix} \beta_{j1} \\ \beta_{j2} \end{pmatrix} = \boldsymbol{\Sigma}^{-1}(\boldsymbol{\mu}_j - \boldsymbol{\mu}_m).$$

As before, $\boldsymbol{\beta}_j = \mathbf{0}$ implies $\alpha_j = 0$, $j = 1, ..., q$, and that $\boldsymbol{\beta}_1 = \cdots = \boldsymbol{\beta}_q = \mathbf{0}$ implies equidistribution: all the $g_i$ are equal.

It is interesting to observe that the density ratio (3.1) has been used for many years in the theory and practice of classification (cf. Anderson 1984, p. 204). For our purposes, however, (3.1) is viewed only as a stepping stone toward a model which has the form (3.2). Consequently, by the *two dimensional density ratio model* we mean the model

$$\frac{g_j(\boldsymbol{x})}{g(\boldsymbol{x})} = \exp(\alpha_j + \boldsymbol{\beta}_j'\boldsymbol{x}), \quad j = 1, ..., q \tag{3.3}$$

with reference $g \equiv g_m$, scalar $\alpha_j$, two-dimensional $\boldsymbol{\beta}_j = (\beta_{j1}, \beta_{j2})'$, and $\boldsymbol{x} = (x, y)'$.

**Remark:** As before, a generalization of (3.3) can be obtained by replacing $\boldsymbol{x}$ in the tilt by a function $\boldsymbol{h}(\boldsymbol{x})$ and modifying $\boldsymbol{\beta}_j$ as needed. For example, taking after the ratio of two bivariate normal densities with unequal covariance matrices, we could use the model

$$\frac{g_j(x, y)}{g_m(x, y)} = \exp(\alpha_j + \boldsymbol{\beta}_j'\boldsymbol{h}(\boldsymbol{x})) \tag{3.4}$$

where

$$\boldsymbol{\beta}_j = (\beta_{j1}, \beta_{j2}, \beta_{j3}, \beta_{j4}, \beta_{j5})' \quad \text{and} \quad \boldsymbol{h}(\boldsymbol{x}) = (x^2, x, y^2, y, xy)'.$$

Clearly, from inferential point of view, the two models (3.3) and (3.4) are essentially the same.

The previous results carry over to the two-dimensional case quite readily. We begin by first defining the combined data,

$$\begin{aligned} \boldsymbol{t} &= (\boldsymbol{x}_{11}', ..., \boldsymbol{x}_{1n_1}', \boldsymbol{x}_{21}', ..., \boldsymbol{x}_{1n_2}', ..., \boldsymbol{x}_{q1}', ..., \boldsymbol{x}_{qn_q}', \boldsymbol{x}_{m1}', ..., \boldsymbol{x}_{mn_m}')' \\ &= (\boldsymbol{t}_1', \boldsymbol{t}_2', \cdots, \boldsymbol{t}_n')'. \end{aligned} \tag{3.5}$$

where $\boldsymbol{t}_i = (t_{ix}, t_{iy})'$. It is convenient to switch between the $\boldsymbol{t}_i$ and $\boldsymbol{x}_{kl}$ as needed.

To obtain the maximum likelihood estimator of $G(x, y)$, we optimize over the class of two-dimensional step functions with jumps $p_i$ at $\boldsymbol{t}_1, ..., \boldsymbol{t}_n$,

$$p_i = G(t_{ix}, t_{iy}) - G(t_{i-1,x}, t_{iy}) - G(t_{ix}, t_{i-1,y}) + G(t_{i-1,x}, t_{i-1,y}), \quad i = 1, ..., n.$$

Defining $\boldsymbol{\alpha} = (\alpha_1, ..., \alpha_q)'$, and $\boldsymbol{\beta} = (\boldsymbol{\beta}_1', ..., \boldsymbol{\beta}_q')'$, the empirical likelihood under model (3.3) is then given by,

$$L(\boldsymbol{\alpha}, \boldsymbol{\beta}, G) = \prod_{i=1}^{n} p_i \prod_{k=1}^{n_1} \exp(\alpha_1 + \beta_{11}x_{1k} + \beta_{12}y_{1k}) \cdots \prod_{k=1}^{n_q} \exp(\alpha_q + \beta_{q1}x_{qk} + \beta_{q2}y_{qk}). \tag{3.6}$$

Subject to the constraints

$$\sum_{i=1}^{n} p_i = 1, \quad \sum_{i=1}^{n} w_1(\boldsymbol{t}_i)p_i = 1, \ldots, \quad \sum_{i=1}^{n} w_q(\boldsymbol{t}_i)p_i = 1 \tag{3.7}$$

where

$$w_j(\boldsymbol{t}_i) = \exp(\alpha_j + \boldsymbol{\beta}_j'\boldsymbol{t}_i) = \exp(\alpha_j + \beta_{j1}t_{ix} + \beta_{j2}t_{iy}), \quad j = 1, ..., q,$$

we obtain the profile log-likelihood

$$\begin{aligned}
\ell(\boldsymbol{\alpha}, \boldsymbol{\beta}) &\equiv \log L(\boldsymbol{\alpha}, \boldsymbol{\beta}, G) \\
&= -\sum_{i=1}^{n} \log[1+\rho_1 w_1(\boldsymbol{t}_i)+\cdots+\rho_q w_q(\boldsymbol{t}_i)] + \sum_{j=1}^{n_1}(\alpha_1+\beta_{11}x_{1j}+\beta_{12}y_{1j}) \\
&\quad +\cdots+\sum_{j=1}^{n_q}(\alpha_q+\beta_{q1}x_{qj}+\beta_{q2}y_{qj}) + \text{Constant.}
\end{aligned} \tag{3.8}$$

Maximizing the log-likelihood (3.8) we obtain the score equations for the maximum likelihood estimators $\hat{\alpha_j}$ and $\hat{\boldsymbol{\beta}}_j$, $j = 1, ..., q$,

$$\frac{\partial \ell}{\partial \alpha_j} = -\sum_{i=1}^{n} \frac{\rho_j w_j(\boldsymbol{t}_i)}{1 + \rho_1 w_1(\boldsymbol{t}_i) + \cdots + \rho_q w_q(\boldsymbol{t}_i)} + n_j = 0 \tag{3.9}$$

$$\frac{\partial \ell}{\partial \boldsymbol{\beta}_j} = -\sum_{i=1}^{n} \frac{\rho_j w_j(\boldsymbol{t}_i)\boldsymbol{t}_i}{1 + \rho_1 w_1(\boldsymbol{t}_i) + \cdots + \rho_q w_q(\boldsymbol{t}_i)} + \sum_{i=1}^{n_j} \begin{pmatrix} x_{ji} \\ y_{ji} \end{pmatrix} = \mathbf{0}, \tag{3.10}$$

and as before,

$$\hat{p}_i = \frac{1}{n_m} \cdot \frac{1}{1 + \rho_1 \hat{w}_1(\boldsymbol{t}_i) + \cdots + \rho_q \hat{w}_q(\boldsymbol{t}_i)} \tag{3.11}$$

$$\hat{G}(\boldsymbol{a}) = \frac{1}{n_m} \cdot \sum_{i=1}^{n} \frac{I_{(-\infty,\boldsymbol{a}]}(\boldsymbol{t}_i)}{1 + \rho_1 \hat{w}_1(\boldsymbol{t}_i) + ... + \rho_q \hat{w}_q(\boldsymbol{t}_i)} \tag{3.12}$$

where $\hat{w}_j(\boldsymbol{t}_i) = \exp(\hat{\alpha}_j + \hat{\boldsymbol{\beta}}_j'\boldsymbol{t}_i)$, and $(-\infty, \mathbf{a}] = (-\infty, a_1] \times (-\infty, a_2]$ for $\mathbf{a} = (a_1, a_2)$. Note that $I_A(\omega) = 1$ for $\omega \in A$ and $I_A(\omega) = 0$ otherwise. It can be shown that the maximum likelihood estimators $\hat{\alpha_j}$ and $\hat{\boldsymbol{\beta}}_j$, $j = 1, ..., q$, are asymptotically normal with a covariance matrix which is essentially the same as the one given in the appendix of Chapter 2.

The equidistribution hypothesis $H_0 : \boldsymbol{\beta}_1 = \boldsymbol{\beta}_2 = \cdots = \boldsymbol{\beta}_q = \mathbf{0}$ can be tested by means of the likelihood ratio (LR),

$$
\begin{aligned}
LR &\equiv -2[\ell(\mathbf{0},\mathbf{0}) - \ell(\hat{\boldsymbol{\alpha}}, \hat{\boldsymbol{\beta}})] \\
&= -2\sum_{i=1}^{n} \log[1 + \rho_1 \hat{w}_1(\boldsymbol{t}_i) + ... + \rho_q \hat{w}_q(\boldsymbol{t}_i)] \\
&+ 2\sum_{i=1}^{q}\sum_{j=1}^{n_i}[\hat{\alpha}_i + \hat{\beta}_{i1}x_{ij} + \hat{\beta}_{i2}y_{ij}] + 2n\log[1 + \sum_{i=1}^{q}\rho_i]. \qquad (3.13)
\end{aligned}
$$

Under $H_0$, the likelihood ratio is approximately distributed as $\chi^2$ with $2q$ degrees of freedom, and $H_0$ is rejected for large values.

### 3.2.1 Case-Control Application

Testicular germ cell tumor (TGCT) is a common cancer among young US men, mainly in the age group 15-35 years (Devesa et al 2003, McGlynn et al 2007). Using odds ratios derived from logistic regression with categories for body size, it was determined in McGlynn et al (2007) that increased height was significantly related to risk. On the other hand, body mass index (weight in kilograms divided by height in meters squared) was not a significant risk factor. It is interesting to find out whether height and weight *jointly* are risk factors, a problem considered in Kedem et al (2009).

We shall apply the two-dimensional semiparametric paradigm discussed above. Such a bivariate analysis can shed light on risk factors which cannot be identified from marginal considerations alone.

After removing incomplete observations, the TGCT data from the Servicemen's Testicular Tumor Environmental and Endocrine Determinants (STEED) (2002-2005) study consist of pairs of heights (cm) and weights (kg) of 1691 individuals, of which $n_1 = 763$ are cases and $n_2 = 928$ are in the control group (McGlynn et al 2007). Summary statistics for both groups are given in Table 3.1 from which the variance-covariance structure in the two groups is quite similar. This supports the density ratio model (3.3) which in the present case is written as,

$$\frac{g_1(x, y)}{g_2(x, y)} = \exp(\alpha_1 + \boldsymbol{\beta}_1' \mathbf{x}) \qquad (3.14)$$

where $g_1$ is the distribution of the case group, and $g_2$ is the reference distribution of the control group, and the hypothesis of interest is that there is no difference bewteen the two groups, that is $H_0 : \boldsymbol{\beta}_1 = \mathbf{0}$.

Table 3.1: Case-control summary statistics regarding height (cm) and weight (kg), and the correlation between them.

| | Height | | | | Weight | | | | |
|---|---|---|---|---|---|---|---|---|---|
| | min | max | ave | sd | min | max | ave | sd | corr |
| Case | 160.0 | 203.2 | 179.6 | 7.0 | 50.8 | 131.5 | 81.4 | 11.7 | 0.521 |
| Control | 152.4 | 215.9 | 178.3 | 7.1 | 38.6 | 127.0 | 80.1 | 11.1 | 0.505 |

From the score equations (3.9) and (3.10),

$$(\hat{\alpha}, \hat{\beta}_{11}, \hat{\beta}_{12}) = (-4.676, 0.025, 0.002) \tag{3.15}$$

with respective standard errors (0.914, 0.006, 0.004), indicating dissimilarity between the two groups. Indeed, the corresponding likelihood ratio test (3.13) with 2 degrees of freedom gives a $p$-value of 0.0005. Thus, when height and weight are considered jointly, we reject the null hypothesis $H_0 : \boldsymbol{\beta}_1 = \mathbf{0}$ of equidistribution quite conclusively, echoing the conclusion reached from height alone.

As remarked earlier, a useful feature of the joint analysis is that it provides estimates of joint probabilities from both case and control groups as shown in Table 3.2. The table points to moderate two-dimensional differences between the two groups, lending support to the testing result.

As a check, it is interesting to see what the two-dimensional analysis gives for "case-case," when the control data are replaced by the case data. Then the coefficients are close to zero,

$$(\hat{\alpha}, \hat{\beta}_{11}, \hat{\beta}_{12}) = (6.415 \times 10^{-06}, -4.870 \times 10^{-08}, 2.919 \times 10^{-08}) \tag{3.16}$$

and the likelihood ratio (3.13) gives a $p$-value close to 1. Also, as expected, when the case and control groups are interchanged the analysis results in the same coefficients but with reversed signs (4.676,-0.025,-0.002).

It is interesting to see what the one-dimensional analog of (3.14) gives when applied to height alone and separately also to body mass index. Thus, in the one-dimensional analysis of height and body mass index, with $h(x) = x$ the hypothesis of equidistribution reduces to testing the vanishing of the scalar $\beta_1$, $\mathrm{H}_0 : \beta_1 = 0$, with $n_1 = 763$ and $n_2 = 928$. Regarding height, the likelihood ratio test (2.38) with one degree of freedom gives a $p$-value of 0.00011, whereas it is 0.88213 for body mass index. Similarly, the $\mathcal{X}_1$

Table 3.2: Some joint probabilities of height (H) and weight (W) in the case and control groups.

| Probability | Case | Control |
|---|---|---|
| $\Pr(H \le 155, W \le 59)$ | 0.000374 | 0.000769 |
| $\Pr(H \le 165, W \le 59)$ | 0.003490 | 0.005750 |
| $\Pr(H \le 178, W \le 65)$ | 0.051604 | 0.066406 |
| $\Pr(H \le 185, W \le 70)$ | 0.133651 | 0.161664 |
| $\Pr(H \le 180, W \le 80)$ | 0.315808 | 0.375041 |
| $\Pr(H \le 180, W \le 90)$ | 0.448033 | 0.520636 |
| $\Pr(H \le 187, W \le 95)$ | 0.769016 | 0.818147 |
| $\Pr(H \le 200, W \le 100)$ | 0.943891 | 0.957770 |
| $\Pr(H \le 203, W \le 119)$ | 0.993959 | 0.996346 |

test (2.27) also with one degree of freedom gives a $p$-values of 0.00009 and 0.88208, respectively. Very similar results are obtained with $h(x) = (x, x^2)$, in which case the tests are with two degrees of freedom. Curiously, the same analysis applied to weight only produces inconclusive or borderline results with the likelihood ratio test and the $\mathcal{X}_1$ tests giving $p$-values of 0.051911 and 0.07043787, respectively.

Consequently, height is a significant risk factor as are height and weight jointly, however, body mass index is not significant in agreement with McGlynn et al (2007) who approached the problem very differently using the odds ratio derived from logistic regression analysis. Concerning weight alone, the results are not quite conclusive.

The discussion in Section 1.1 concerning logistic regression is very relevant here. As noted there, $\boldsymbol{\beta}$ can be estimated using the logistic model (1.5), however, since we have case-control data, the density ratio model (3.14) provides *in addition* inference about the case and control *distributions* $g_1(x, y)$ and $g_2(x, y)$, respectively. Kernel density estimation is discussed in the next section.

## 3.3 The General Multivariate Case

The extension of the bivariate density ratio model (3.3) to the general multivariate case requires little effort. The extension is motivated by numerous problems when there are a number of multidimensional samples such as those encountered in case-control studies where measurements are available on several variables, for example age, height, and weight. An inspiring example is the problem of general regression with random covariates where the data consist of multivariate vectors of which one component is a response and the rest are random covariates, a problem discussed in the next section.

So, suppose we have $m = q + 1$ independent $p$-dimensional random samples

$$\begin{gathered}
\boldsymbol{x}_{11}, \ldots, \boldsymbol{x}_{1n_1} \overset{iid}{\sim} g_1(\boldsymbol{x}) \\
\vdots \\
\boldsymbol{x}_{q1}, \ldots, \boldsymbol{x}_{qn_q} \overset{iid}{\sim} g_q(\boldsymbol{x}) \\
\boldsymbol{x}_{m1}, \ldots, \boldsymbol{x}_{mn_m} \overset{iid}{\sim} g_m(\boldsymbol{x})
\end{gathered}$$

where, for $i = 1, \ldots, q, m,\ j = 1, \ldots, n_i$,

$$\boldsymbol{x}_{ij} = (x_{ij1}, x_{ij2}, \ldots, x_{ijp}) \sim g_i(x_1, \ldots, x_p)$$

and $g_i(x_1, x_2, \ldots, x_p)$ is the probability density function (pdf) corresponding to the $i$th sample. As before, the $i$th sample size is $n_i$, $n = \sum_{i=1}^{m} n_i$ is the total sample size, and $g \equiv g_m(\boldsymbol{x}) \equiv g_m(x_1, \ldots, x_p)$ is the reference or baseline probability density function.

Extending the bivariate case, we assume the density ratio model with an exponential tilt parametrized by $\boldsymbol{\beta}_i = (\beta_{i1}, \ldots, \beta_{ip})'$,

$$\frac{g_i(\boldsymbol{x})}{g(\boldsymbol{x})} = w_i(\boldsymbol{x}) \equiv \exp(\alpha_i + \boldsymbol{\beta}_i'\boldsymbol{x}), \quad i = 1, ..., q. \tag{3.17}$$

Since the $g_i(\boldsymbol{x})$ are probability densities, $\boldsymbol{\beta}_i = \mathbf{0}$ implies $\alpha_i = 0$, and $\boldsymbol{\beta}_1 = \cdots = \boldsymbol{\beta}_q = \mathbf{0}$ implies equidistribution: all the $g_i$ are equal.

As before, $G(\boldsymbol{x}) \equiv G_m(\boldsymbol{x})$ is the reference cdf and we define the jumps $p_{ij} = dG(\boldsymbol{x}_{ij})$. The empirical likelihood from the pooled or fused data $\boldsymbol{x}_{ij}$, $i = 1, \ldots, m$, $j = 1, \ldots, n_i$, is given by

$$\begin{aligned}
L(\boldsymbol{\alpha}, \boldsymbol{\beta}, G) &= \left[\prod_{j=1}^{n_1} p_{1j} w_1(\boldsymbol{x}_{1j})\right] \left[\prod_{j=1}^{n_2} p_{2j} w_2(\boldsymbol{x}_{2j})\right] \cdots \left[\prod_{j=1}^{n_m} p_{mj}\right] \\
&= \left[\prod_{i=1}^{m} \prod_{j=1}^{n_i} p_{ij}\right] \left[\prod_{i=1}^{q} \prod_{j=1}^{n_i} w_i(\boldsymbol{x}_{ij})\right]. \qquad (3.18)
\end{aligned}$$

The log-likelihood is then,

$$\ell = \log L = \sum_{i=1}^{m}\sum_{j=1}^{n_i} \log(p_{ij}) + \sum_{i=1}^{q}\sum_{j=1}^{n_i} \log(w_i(\boldsymbol{x}_{ij})) \tag{3.19}$$

with constraints,

$$p_{ij} \geq 0, \sum_{i=1}^{m}\sum_{j=1}^{n_i} p_{ij} = 1, \sum_{i=1}^{m}\sum_{j=1}^{n_i} p_{ij} w_k(\boldsymbol{x}_{ij}) = 1, \quad k = 1, \ldots, q. \tag{3.20}$$

Fokianos (2004), and Qin and Lawless (1994) gave conditions guaranteeing that, with probability approaching 1, there is a maximum in a small neighborhood of the true parameters $\boldsymbol{\alpha}_0, \boldsymbol{\beta}_0$.

Observe that $\boldsymbol{x}_{ij}$ corresponds to some point $\boldsymbol{t}_i$ in the combined data $\boldsymbol{t}$, a vector of length $n = n_1 + n_2 + \cdots + n_m$ with components $\boldsymbol{t}_i$ as in (3.5). It is convenient at times to switch notation from the $\boldsymbol{x}_{ij}$ to its corresponding point $\boldsymbol{t}_i$. The advantage in the notation $\boldsymbol{x}_{ij}$ is that it refers to a specific vector in the $i$th sample, whereas this information is lost when using the corresponding $\boldsymbol{t}_i$ instead. However, the later notation, that is $\boldsymbol{t}_i$, is simpler and avoids double summation.

Then, as was done earlier, maximizing the log-likelihood (3.19) subjects to the constraints (3.20) we obtain the score equations for $\hat{\alpha}_j$ and $\hat{\boldsymbol{\beta}}_j$, $j = 1, \ldots, q$,

$$\frac{\partial \ell}{\partial \alpha_j} = - \sum_{i=1}^{n} \frac{\rho_j w_j(\boldsymbol{t}_i)}{1 + \rho_1 w_1(\boldsymbol{t}_i) + \cdots + \rho_q w_q(\boldsymbol{t}_i)} + n_j = 0 \tag{3.21}$$

$$\begin{aligned} \frac{\partial \ell}{\partial \boldsymbol{\beta}_j} &= - \sum_{i=1}^{n} \frac{\rho_j w_j(\boldsymbol{t}_i)\boldsymbol{t}_i}{1 + \rho_1 w_1(\boldsymbol{t}_i) + \cdots + \rho_q w_q(\boldsymbol{t}_i)} \\ &+ \sum_{i=1}^{n_j} (x_{ji1}, \ldots, x_{jip})' = \mathbf{0} \end{aligned} \tag{3.22}$$

where $\rho_j = n_j/n_m$ are the relative sample sizes. Following Lu (2007) it can be shown that as $n \to \infty$ the $\hat{\alpha}_i$ and $\hat{\boldsymbol{\beta}}_i$ are asymptotically normal. As in the univariate case, substituting the $\hat{\alpha}_i$ and $\hat{\boldsymbol{\beta}}_i$ we obtain the familiar expressions,

$$\hat{p}_i = \frac{1}{n_m} \cdot \frac{1}{1 + \rho_1 \hat{w}_1(\boldsymbol{t}_i) + \cdots + \rho_q \hat{w}_q(\boldsymbol{t}_i)} \tag{3.23}$$

$$\hat{G}(\boldsymbol{x}) = \frac{1}{n_m} \cdot \sum_{i=1}^{n} \frac{I(\boldsymbol{t}_i \leq \boldsymbol{x})}{1 + \rho_1 \hat{w}_1(\boldsymbol{t}_i) + \cdots + \rho_q \hat{w}_q(\boldsymbol{t}_i)} \tag{3.24}$$

where $(\boldsymbol{t}_i \leq \boldsymbol{x})$ is defined componentwise, $\hat{w}_j(\boldsymbol{t}_i) = \exp(\hat{\alpha}_j + \hat{\boldsymbol{\beta}}_j' \boldsymbol{t}_i)$, and $I(B)$ is the indicator of the event $B$.

In terms of $\boldsymbol{x}_{ij}$ then,

$$\hat{p}_{ij} = \frac{1}{n_m} \cdot \frac{1}{1 + \rho_1 \hat{w}_1(\boldsymbol{x}_{ij}) + \cdots + \rho_q \hat{w}_q(\boldsymbol{x}_{ij})} \tag{3.25}$$

$$\hat{G}_m(\boldsymbol{x}) = \hat{G}(\boldsymbol{x}) = \sum_{i=1}^{m} \sum_{j=1}^{n_i} \hat{p}_{ij} I(\mathbf{x}_{ij} \leq \mathbf{x}) \tag{3.26}$$

$$\hat{G}_k(\boldsymbol{x}) = \sum_{i=1}^{m} \sum_{j=1}^{n_i} \hat{p}_{ij} w_k(\boldsymbol{x}_{ij}) I(\boldsymbol{x}_{ij} \leq \boldsymbol{x}) \tag{3.27}$$

### 3.3.1 Kernel Density Estimation Using Fused Data

One of the useful byproducts of the density ratio model is that it allows the estimation of the reference $g$ and all the $g_l$ from the combined data by operating or smoothing the $\hat{p}_{ij}$ and their distortions $\hat{p}_{ij}\hat{w}_l(\boldsymbol{x}_{ij})$ by means of a kernel function. When the density ratio model (3.17) holds, then the fused estimate is more precise than its traditional kernel estimate counterpart obtained only from a single sample.

The idea of constructing kernel density estimates by smoothing the increments of $\hat{G}_l$, obtained from the pooled data, rather than smoothing the increments $1/n_l$ in the empirical distribution from the $l$th sample only, has been discussed by a number of authors including, Fokianos (2004), Cheng and Chu (2004), and Qin and Zhang (2005). Most of this work caters to various versions of the univariate case. We follow the extension to the multivariate case discussed in Voulgaraki et al. (2012). The idea can be traced back to the work of Jones (1991) on length bias data and its multivariate extension in Ahmad (1995). The references mentioned here refer further to a voluminous and extensive earlier body of work on kernel density estimation.

By a kernel we mean any function $K(\boldsymbol{x})$ defined for $p$-dimensional vectors $\boldsymbol{x}$, which is nonnegative, symmetric around $\mathbf{0}$, and satisfies the following conditions:

1. $\int K(\boldsymbol{x})d\boldsymbol{x} = 1 \qquad \int |K(\boldsymbol{x})|d\boldsymbol{x} < \infty$
2. $\int \boldsymbol{x}K(\boldsymbol{x})d\boldsymbol{x} = \mathbf{0} \qquad \int |\boldsymbol{x}K(\boldsymbol{x})|d\boldsymbol{x} < \infty$
3. $\int \boldsymbol{x}\boldsymbol{x}'K(\boldsymbol{x})d\boldsymbol{x} = k_2 \qquad \int |\boldsymbol{x}\boldsymbol{x}'K(\boldsymbol{x})|d\boldsymbol{x} < \infty.$

The traditional multivariate kernel density estimator of a probability density

$f(\boldsymbol{x})$ is formulated as,

$$\hat{f}(\boldsymbol{x}) = \frac{1}{nh_n^p}\sum_{i=1}^{n} K\left(\frac{\boldsymbol{x}-\boldsymbol{x}_i}{h_n}\right) \tag{3.28}$$

where $h_n$ is a sequence of bandwidths such that $h_n \to 0$ and $nh_n^p \to \infty$ as $n \to \infty$. The bandwidth parameter controls the smoothness of the kernel estimator: as the bandwidth grows the estimator becomes smoother. Under certain conditions, $\hat{f}(\boldsymbol{x})$ is a consistent estimator of $f(\boldsymbol{x})$ (Rosenblatt 1956, Parzen 1962, Silverman 1986). By its very nature, the traditional kernel density estimator is a "single sample" estimator as opposed to the fused estimator obtain from multiple samples as defined next.

Concerning the $l$th sample, in the fused kernel estimator from the combined data the $1/n_l$ jumps are replaced by the tilted jumps $\hat{p}_{ij}\hat{w}_l(\boldsymbol{x}_{ij})$, with $\hat{p}_{ij}$ given in (3.25), to form the kernel density estimator of the probability density $g_l(\mathbf{x})$ (Fokianos 2004, Qin and Zhang 2005),

$$\hat{g}_l(\mathbf{x}) = \frac{1}{h_n^p}\sum_{i=1}^{m}\sum_{j=1}^{n_i} \hat{p}_{ij}\hat{w}_l(\boldsymbol{x}_{ij}) K\left(\frac{\boldsymbol{x}-\boldsymbol{x}_{ij}}{h_n}\right) \tag{3.29}$$

where again $h_n$ is a sequence of bandwidths such that $h_n \to 0$ and $nh_n^p \to \infty$, as $n \to \infty$. Observe that $\hat{g}_l$ is a proper probability density function. The use of (3.29) is illustrated in a number of the previous figures including Figures 2.5 to 2.8.

Assuming the density ratio model (3.17), we shall describe next only certain useful facts concerning the asymptotic behavior of $\hat{g}_l$ and refer the reader to Fokianos (2004), Qin and Zhang (2005), and Voulgaraki et al. (2012) for complete proofs and technical details.

First, it is helpful to define,

$$\tilde{g}_l(\boldsymbol{x}) = \frac{1}{h_n^p}\sum_{i=1}^{m}\sum_{j=1}^{n_i} p_{ij} w_l(\boldsymbol{x}_{ij}) K\left(\frac{\boldsymbol{x}-\boldsymbol{x}_{ij}}{h_n}\right). \tag{3.30}$$

Asymptotically $\hat{g}_l(\boldsymbol{x})$ and $\tilde{g}_l(\boldsymbol{x})$ are not far apart, however, it is significantly simpler to handle the latter to shed light on some general properties shared by the two estimators.

Assume that $K(\cdot)$ is a nonnegative bounded symmetric kernel and that $g_l$ is twice continuously differentiable in a neighborhood of $\boldsymbol{x}$, then as $n \to \infty$, $h_n \to 0$, and $nh_n^p \to \infty$,

$$\mathrm{E}(\tilde{g}_l(\boldsymbol{x})) = g_l(\boldsymbol{x}) + \frac{1}{2}h_n^2 \int \boldsymbol{u}' \frac{\partial^2}{\partial\boldsymbol{x}\partial\boldsymbol{x}'} g_l(\boldsymbol{x})\boldsymbol{u}K(\boldsymbol{u})d\boldsymbol{u} + o(h_n^2), \tag{3.31}$$

and

$$\mathrm{Var}(\tilde{g}_l(\boldsymbol{x})) = \frac{1}{nh_n^p}\sigma_l^2(\boldsymbol{x}) + o\left(\frac{1}{nh_n^p}\right) \tag{3.32}$$

where

$$\sigma_l^2(\boldsymbol{x}) = \frac{w_l(\boldsymbol{x})g_l(\boldsymbol{x})}{\sum_{k=1}^m \zeta_k w_k(\boldsymbol{x})}\int K^2(\boldsymbol{u})d\boldsymbol{u}$$

for any fixed $\boldsymbol{x}$, and $\zeta_k$ are positive limits of $n_k/n$, $k = 1, .., q$. With this, after some tedious work, it can be shown that if as $n \to \infty$, $h_n = O(n^{-\frac{1}{4+p}})$, then $\hat{g}_l$ is asymptotically normal,

$$\sqrt{nh_n^p}\left(\hat{g}_l(\boldsymbol{x}) - g_l(\boldsymbol{x}) - \frac{1}{2}h_n^2\int \boldsymbol{u}'\frac{\partial^2 g_l(\boldsymbol{x}^*)}{\partial\boldsymbol{x}\partial\boldsymbol{x}'}\boldsymbol{u}K(\boldsymbol{u})d\boldsymbol{u}\right) \xrightarrow{D} N(\boldsymbol{0}, \sigma_l^2(\boldsymbol{x})). \tag{3.33}$$

An important lesson to learn from (3.31)-(3.33) is that there is a tradeoff between bias and variance as a function of bandwidth. A small bandwidth results in a small bias but large variance and vice-versa. This bias-variance tradeoff is a perpetual dilemma encountered also in traditional kernel density estimation and in spectral density estimation. A way out is to obtain a bandwidth which is optimal in some sense. To this end we define a measure of precision to be minimized as a function of bandwidth.

The precision of $\hat{g}_l$ can be measured by the *mean integrated square error* (MISE)

$$\mathrm{MISE}(\hat{g}_l) = \mathrm{E}\int |\hat{g}_l(\boldsymbol{x}) - g_l(\boldsymbol{x})|^2 d\boldsymbol{x}.$$

We can use the MISE to compare the traditional and fused kernel estimators. Let

$$\hat{f}_l(\mathbf{x}) = \frac{1}{n_l h_n^p}\sum_{i=1}^{n_l} K\left(\frac{\mathbf{x} - \mathbf{x}_i}{h_n}\right)$$

denote the traditional single-sample multivariate kernel density estimator of $g_l$. As $n \to \infty$, $h_n \to 0$ and $nh_n^p \to \infty$ (Cacoullos 1966), the *asymptotic mean integrated square error* (AMISE) from $\hat{f}_l$ is,

$$\mathrm{AMISE}(\hat{f}_l) = \frac{1}{4}h_n^4\int\left(\int \boldsymbol{u}'\frac{\partial^2 g_l(\boldsymbol{x})}{\partial\boldsymbol{x}\partial\boldsymbol{x}'}\boldsymbol{u}K(\boldsymbol{u})d\boldsymbol{u}\right)^2 d\boldsymbol{x} + \frac{1}{n_l h_n^p}\int K^2(\boldsymbol{x})d\boldsymbol{x}.$$

Similarly, it can be shown that (Fokianos 2004, Voulgarki et al. 2012),

$$\text{AMISE}(\hat{g}_l) = \frac{1}{4}h_n^4 \int \left( \int \boldsymbol{u}' \frac{\partial^2 g_l(\boldsymbol{x})}{\partial \boldsymbol{x} \partial \boldsymbol{x}'} \boldsymbol{u} K(\boldsymbol{u}) d\boldsymbol{u} \right)^2 d\boldsymbol{x}$$
$$+ \frac{1}{n_l h_n^p} \int \frac{\zeta_l w_l(\boldsymbol{x}) g_l(\boldsymbol{x})}{\sum_{k=1}^m \zeta_k w_k(\boldsymbol{x})} d\boldsymbol{x} \int K^2(\boldsymbol{x}) d\boldsymbol{x}.$$

Since for every $l$,

$$\int \frac{\zeta_l w_l(\boldsymbol{x}) g_l(\boldsymbol{x})}{\sum_{k=1}^m \zeta_k w_k(\boldsymbol{x})} d\boldsymbol{x} \le 1,$$

it follows that asymptotically,

$$\text{AMISE}(\hat{g}_l) \le \text{AMISE}(\hat{f}_l).$$

Thus, asymptotically, when the density ratio model holds the fused estimator dominates the traditional single sample kernel estimator, which is expected as the fused estimator is obtained from more data.

### 3.3.2 Bandwidth selection for $\hat{g}_l$

As in the classical single sample kernel density estimtor, an optimal bandwidth can be obtained by minimizing MISE($\hat{g}_l$). In the Supplement to Voulgaraki et al. (2012) it is shown that the optimal bandwidth is given by the expression,

$$h_n^* = \left( \frac{(p/n) \int w_l(\boldsymbol{x}) g_l(\boldsymbol{x}) / \left[\sum_{k=1}^m \zeta_k w_k(\boldsymbol{x})\right] d\boldsymbol{x} \int K^2(\boldsymbol{u}) d\boldsymbol{u}}{\int \left( \int \boldsymbol{u}'(\partial^2 g_l(\boldsymbol{x})/\partial \boldsymbol{x} \partial \boldsymbol{x}') \boldsymbol{u} K(\boldsymbol{u}) d\boldsymbol{u} \right)^2 d\boldsymbol{x}} \right)^{\frac{1}{4+p}} \tag{3.34}$$

which is somewhat lacking as it depends on the unknown $g_l$ and its second order partial derivatives. In the one dimensional case, as some authors suggest, $g_l$ can be replaced by $\hat{g}_l$ or by the normal density from $\text{N}(\mu, \sigma^2)$ where $\mu$ and $\sigma^2$ are estimated from the $l$th sample. In the multivariate version these remedies are less appealing in light of the partial derivatives. We discuss next practical ways for choosing bandwidths which are optimal in some sense.

A practical way out is to modify the objective function and minimize with respect to the bandwidth $h_n$ the *integrated squared error* (ISE),

$$\begin{aligned} \text{ISE}(h_n) &= \int (\hat{g}_l(\boldsymbol{x}) - g_l(\boldsymbol{x}))^2 d\boldsymbol{x} \\ &= \int \hat{g}_l^2(\boldsymbol{x}) d\boldsymbol{x} - 2 \int \hat{g}_l(\boldsymbol{x}) g_l(\boldsymbol{x}) d\boldsymbol{x} + \int g_l^2(\boldsymbol{x}) d\boldsymbol{x}. \end{aligned}$$

As the last term does not depend on $h_n$ it can be ignored so that to minimize the ISE we need to consider the first and second terms only. The first term can be expressed after a change of variable as

$$\int \hat{g}_l^2(\boldsymbol{x})d\boldsymbol{x} = \int \left[\frac{1}{h_n^p}\sum_{i=1}^{m}\sum_{j=1}^{n_i} \hat{p}_{ij}\hat{w}_l(\boldsymbol{x}_{ij})K\left(\frac{\boldsymbol{x}-\boldsymbol{x}_{ij}}{h_n}\right)\right]^2 d\boldsymbol{x}$$

$$= h_n^{-p}\sum_{i=1}^{n}\sum_{i'=1}^{n} \hat{p}(\boldsymbol{t}_i)\hat{w}_l(\boldsymbol{t}_i)\hat{p}(\boldsymbol{t}_{i'})\hat{w}_l(\boldsymbol{t}_{i'}) \int K(\boldsymbol{z})K\left(\boldsymbol{z}+\frac{\boldsymbol{t}_i-\boldsymbol{t}_{i'}}{h_n}\right)d\boldsymbol{z}.$$

As for the second term we observe that $\int \hat{g}_l(\boldsymbol{x})g_l(\boldsymbol{x})d\boldsymbol{x} = \mathrm{E}(\hat{g}_l(\boldsymbol{x}))$. Following the cross-validation method of Silverman (1986), $\mathrm{E}(\hat{g}_l(\boldsymbol{x}))$ can be estimated by the *leave one out estimator*,

$$\widehat{\mathrm{E}\hat{g}_l(\boldsymbol{x})} = \frac{1}{n_l}\sum_{i=n_1+\ldots+n_{l-1}+1}^{n_l} \hat{g}_{l,-i}(\boldsymbol{t}_i)$$

where $\hat{g}_{l,-i}(\boldsymbol{t}_i)$ is $\hat{g}_l(\boldsymbol{t}_i)$ with $\boldsymbol{t}_i$ dropped from the combined data. Therefore, a nearly optimal bandwidth $h_n$ minimizes the objective function

$$h_n^{-p}\sum_{i=1}^{n}\sum_{i'=1}^{n} \hat{p}(\boldsymbol{t}_i)\hat{w}_l(\boldsymbol{t}_i)\hat{p}(\boldsymbol{t}_{i'})\hat{w}_l(\boldsymbol{t}_{i'}) \int K(\boldsymbol{z})K\left(\boldsymbol{z}+\frac{\boldsymbol{t}_i-\boldsymbol{t}_{i'}}{h_n}\right)d\boldsymbol{z}$$
$$-\frac{2}{n_l}\sum_{i=n_1+\ldots+n_{l-1}+1}^{n_l} \hat{g}_{l,-i}(\boldsymbol{t}_i). \quad (3.35)$$

An alternative method which is computationally simpler but requires large samples uses the approximation

$$\int \hat{g}_l(\boldsymbol{x})g_l(\boldsymbol{x})d\boldsymbol{x} \approx \int \tilde{g}_l(\boldsymbol{x})g_l(\boldsymbol{x})d\boldsymbol{x} \approx E\left[\int \tilde{g}_l(\boldsymbol{x})g_l(\boldsymbol{x})d\boldsymbol{x}\right]$$
$$= \frac{1}{n_l(n_l-1)h_n^p}\sum_{i\neq j} K\left(\frac{\boldsymbol{x}_{li}-\boldsymbol{x}_{lj}}{h_n}\right).$$

Therefore, another way to obtain near optimal bandwidths is to determine $h_n$ which minimizes

$$h_n^{-p}\sum_{i=1}^{n}\sum_{i'=1}^{n} \hat{p}(\boldsymbol{t}_i)\hat{w}_l(\boldsymbol{t}_i)\hat{p}(\boldsymbol{t}_{i'})\hat{w}_l(\boldsymbol{t}_{i'}) \int K(\mathbf{z})K\left(\mathbf{z}+\frac{\boldsymbol{t}_i-\boldsymbol{t}_{i'}}{h_n}\right)d\mathbf{z}$$
$$-\frac{2}{n_l(n_l-1)h_n^p}\sum_{i\neq j} K\left(\frac{\boldsymbol{x}_{li}-\boldsymbol{x}_{lj}}{h_n}\right). \quad (3.36)$$

## 3.4 Semiparametric Regression

Emboldened by the enhanced precision of the multiple-source semiparametric multivariate kernel density estimates discussed in the previous section, we propose, as a byproduct, a semiparametric approach to the estimation of the conditional expectation of a response $y$ given its random covariates $(x_1, ..., x_{p-1})$. This parallels what is known in the literature as linear multiple regression with "random $x$'s", except that the model discussed here is not linear, the normal assumption regarding the joint distribution of the response and the covariates is not made, and the kernel density estimates are obtained by fusing data from $m$ sources. Notice that in our semiparametric setting we actually estimate $m$ conditional expectations corresponding to the $m$ data sources, whereas in classical regression with random covariates there is a single data source and a single regression problem. Thus in case-control problems we get $\mathrm{E}(y|\boldsymbol{x})$ corresponding to both case and control by fusing the data from both sources.

Under the same multivariate setup leading to the density ratio model (3.17), suppose we have $m = q + 1$ samples of $p$-dimensional vectors, where now each vector consists of $p - 1$ covariates and a response,

$$\boldsymbol{x}_{ij} = (x_{ij1}, x_{ij2}, \ldots, x_{ij(p-1)}, y_{ij}) \sim g_i(x_1, \ldots, x_{(p-1)}, y),$$

where $i = 1, \ldots, q, m,\ j = 1, \ldots, n_i$. Let $g \equiv g_m(x_1, \ldots, x_{(p-1)}, y)$ be the reference probability density function (pdf), and assume

$$\frac{g_i(\boldsymbol{x})}{g(\boldsymbol{x})} = \exp(\alpha_i + \boldsymbol{\beta}_i'\boldsymbol{x}), \quad i = 1, ..., q.$$

Then the conditional expectation $\mathrm{E}_i[y|\boldsymbol{x}]$ of source $i$,

$$\mathrm{E}_i[y|\boldsymbol{x}] = \frac{\int y g_i(\boldsymbol{x}, y) dy}{\int g_i(\boldsymbol{x}, y) dy}$$

can be estimated by

$$\hat{\mathrm{E}}_i(y \mid x_1, \ldots, x_{(p-1)}) = \sum_j^{n_i} y_j \frac{\hat{g}_i(x_1, \ldots, x_{(p-1)}, y_j)}{\sum_j \hat{g}_i(x_1, \ldots, x_{(p-1)}, y_j)}, \quad i = 1, \ldots, q, m, \tag{3.37}$$

where the summation is over the $i$th sample. The $\hat{g}_i$ in (3.37) are the semiparametric kernel density estimates discussed in the previous section,

$$\hat{g}_i(\boldsymbol{z}) = \frac{1}{h_n^p} \sum_{k=1}^{n} \hat{p}_k \hat{w}_i(\boldsymbol{t}_k) K\left(\frac{\boldsymbol{t}_k - \boldsymbol{z}}{h_n}\right). \tag{3.38}$$

The consistency of (3.37) is established in Voulgaraki et al. (2012) under the assumption that the data are bounded.

A single sample counterpart of the estimator (3.37) is the well known Nadaraya-Watson nonparametric weighted (local) average (Nadaraya 1964, Watson 1964),

$$\hat{m}_i(\boldsymbol{x}) = \frac{\sum_j^{n_i} K(\boldsymbol{H}^{-1}(\boldsymbol{x}_j - \boldsymbol{x}))y_j}{\sum_j^{n_i} K(\boldsymbol{H}^{-1}(\boldsymbol{x}_j - \boldsymbol{x}))} \tag{3.39}$$

where $\boldsymbol{H}$ is a symmetric *bandwidth matrix* which could be diagonal. Observe that both (3.37) and (3.39) are weighted averages of the form $\sum_i w_i y_i$ where the $w_i$ are positive weights which sum to 1, except that in (3.37) the $w_i$ depend on the $y_i$. Comparisons between both estimators using simulated and real data show that in general both estimators are quite comparable although the Nadaraya-Watson estimator has a slight advantage as judged by mean absolute deviation. This can be explained by the fact that our method consists of an extra step of density estimation.

### 3.4.1 Case-Control Application

We continue the TGCT example from Section 3.2.1 by adding age as an additional covariate. The problem is to estimate $\mathrm{E}(Weight|Height, Age)$ using the estimator (3.37) by fusing 3D vectors of age, height, and weight from case ($n_1 = 763$) and control data ($n_2 = 928$). To check model validity, Figure 3.1 shows plots of $\hat{G}_i$ versus their empirical counterparts $\tilde{G}_i$, $i = 1, 2$, evaluated at selected (age,height,weight) triplets. The pairs ($\hat{G}_i$,$\tilde{G}_i$) lie on 45-degree lines pointing to a a good match in both the case and control groups. In addition, the residual plots in Figure 3.2 point to a reasonable goodness of fit of model (3.17). In evaluating the conditional expectation estimates for both case and control, the optimal bandwidths were 2.24 for control and 2.5 for case, and $K$ was the standard multivariate normal kernel.

Table 3.3 gives a few predicted values of weight given age and height for case and control. From the table we can see that across different ages, except for heights less than 167.64 cm, the predicted weight is consistently slightly greater in case than in control, possibly indicating that increased caloric diet intake or lack of physical exercise may increase the risk of testicular cancer. The much larger complete table is given in Voulgaraki et al. (2012). In the present example the results from multiple regression with random covariates and from the Nadaraya-Watson method are quite similar.

In general, as expected, $\hat{E}(y|\mathbf{x})$ in (3.37) tends to be close to the average of $y$'s which correspond to the same $\mathbf{x}$. This averaging property has been

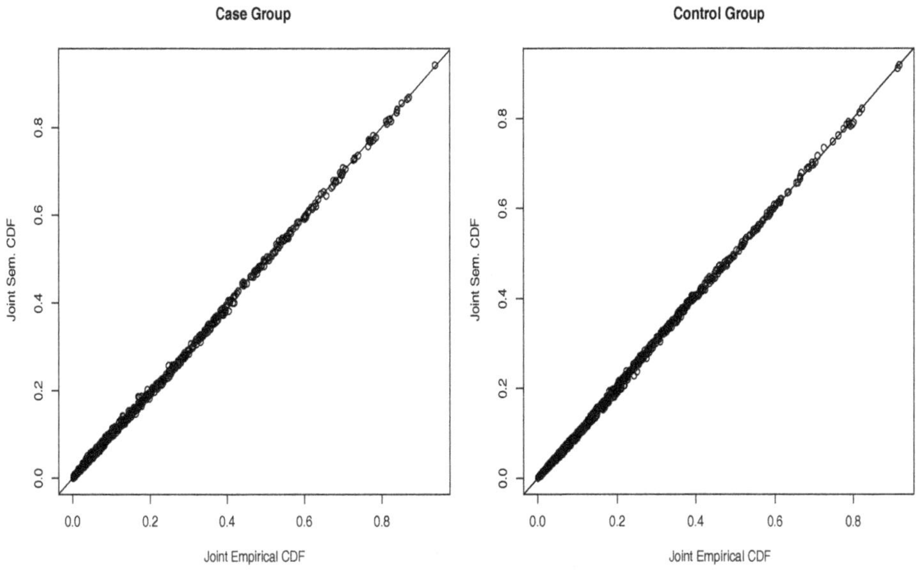


Figure 3.1: Case-control diagnostic plots of $\hat{G}_i$ versus $\tilde{G}_i$, $i = 1, 2$ from 3D TGCT data.

observed in the present example as well. See Voulgaraki et al. (2012) for more details and for examples of joint probabilities obtained from $\hat{G}, \hat{G}_1$ in (3.27).

### 3.4.2 Marginalized Empirical Likelihood

In certain situations the statistician is faced with exceedingly small sample sizes perhaps as small as 1 in which case the combined sample may not give much information about all the parameters, finite and infinite dimensional, in a density ratio model. This is encountered, for example, in statewide surveys when only a few observations are available from some counties, or in nationwide surveys when there are only a few observations from some states. In today's vernacular this might be referred to a "small area" problem. Is there a way to reduce the number of parameters to manage the scant information? Taking a clue from random effects in mixed effects models, the dimensionality of the problem can be reduced considerably by assuming that the $\alpha_i$ and $\boldsymbol{\beta}_i$ are drawn from lower dimensional parametric distributions whose parameters can be estimated without difficulty, and integrating the

Table 3.3: Predicted weight $\hat{\mathrm{E}}$[weight|height, age].

| Height | Age | Control $\hat{\mathrm{E}}[W \mid H, A]$ | Case $\hat{\mathrm{E}}[W \mid H, A]$ |
|---|---|---|---|
| 162.56 | 27 | 69.08335 | 68.53652 |
| 162.56 | 28 | 69.05132 | 68.59858 |
| 165.10 | 30 | 72.20524 | 72.00280 |
| 165.10 | 37 | 72.42138 | 71.85040 |
| 167.64 | 25 | 73.68129 | 73.69978 |
| 167.64 | 30 | 74.81333 | 74.93543 |
| 170.18 | 18 | 73.67032 | 73.67518 |
| 170.18 | 32 | 76.53351 | 76.64543 |
| 172.72 | 37 | 77.88598 | 77.94170 |
| 172.72 | 40 | 77.97789 | 78.04410 |
| 175.26 | 22 | 76.62195 | 76.70862 |
| 175.26 | 25 | 77.14234 | 77.21755 |
| 177.80 | 26 | 78.74752 | 78.92705 |
| 177.80 | 42 | 80.50100 | 80.67493 |
| 180.34 | 20 | 79.17623 | 79.35688 |
| 180.34 | 33 | 81.92536 | 82.17689 |
| 182.88 | 18 | 80.23013 | 80.29011 |
| 182.88 | 41 | 83.65558 | 84.06475 |
| 185.42 | 19 | 81.45580 | 82.09186 |
| 185.42 | 21 | 82.46773 | 82.78140 |
| 190.50 | 22 | 85.23493 | 85.64845 |
| 190.50 | 31 | 86.05980 | 86.27744 |
| 193.04 | 22 | 86.73352 | 87.18440 |
| 193.04 | 24 | 87.50020 | 88.23938 |
| 193.04 | 34 | 87.72937 | 88.58960 |
| 195.58 | 34 | 88.81524 | 89.03654 |

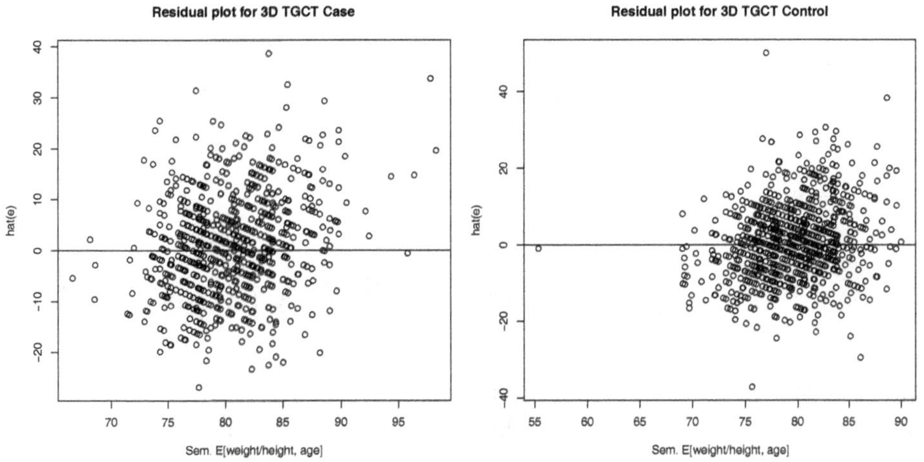


Figure 3.2: Residual plots from (3.37) using the 3D TGCT data.

empirical likelihood against the distributions of the $\alpha_i$ and $\boldsymbol{\beta}_i$. Following a similar reasoning in logistic regression while accommodating heterogeneity in consumer purchasing behavior, this marginalization idea has been developed in Dayaratna (2014) in connection with tort reform analysis as follows.

Assume that in the empirical likelihood (3.18) the parameters are random where $\alpha_i \sim \mathrm{N}(\mu_\alpha, 1)$ and $\boldsymbol{\beta}_i \sim \mathrm{N}(\boldsymbol{\mu}_\beta, \boldsymbol{\Sigma}_\beta)$ are the the so called "heterogeneity distributions". To simplify matters the $\alpha_i$ and $\boldsymbol{\beta}_i$ are assumed independent and $\boldsymbol{\Sigma}_\beta$ is taken as a diagonal matrix so that its components $\beta_{i1}, ..., \beta_{ip}$ are independent as well. As we have in mind applications where there is exactly one observation from each "individual", we assume in what follows that $n_i = 1$, $i = 1, ..., q$.

By integrating the empirical likelihood (3.18) against the heterogeneity distributions we obtain the "marginalized empirical likelihood,"

$$\mathrm{ML}(\mu_\alpha, \boldsymbol{\mu}_\beta, \boldsymbol{\Sigma}_\beta, G) = \prod_{i=1}^{m} \prod_{j=1}^{n_i} p_{ij} \prod_{i=1}^{q} \prod_{j=1}^{n_i} e^{\mu_\alpha + \frac{1}{2}} e^{\boldsymbol{\mu}_\beta' \boldsymbol{x}_{ij} + \frac{1}{2} \boldsymbol{x}_{ij}' \boldsymbol{\Sigma}_\beta^{-1} \boldsymbol{x}_{ij}} \quad (3.40)$$

and the corresponding marginalized log-likelihood,

$$\log \mathrm{ML}(\mu_\alpha, \boldsymbol{\mu}_\beta, \boldsymbol{\Sigma}_\beta, G)$$
$$= \sum_{i=1}^{m}\sum_{j=1}^{n_i} \log p_{ij} + \sum_{i=1}^{q}\sum_{j=1}^{n_i} (\mu_\alpha + \frac{1}{2} + \boldsymbol{\mu}_\beta{}' \boldsymbol{x}_{ij} + \frac{1}{2}\boldsymbol{x}_{ij}{}'\boldsymbol{\Sigma}_\beta^{-1}\boldsymbol{x}_{ij}), \tag{3.41}$$

which is maximized subject to constraints analogous to those in (3.20), $p_{ij} \geq 0$, $\sum_{i=1}^{m}\sum_{j=1}^{n_i} p_{ij} = 1$, and

$$\sum_{i=1}^{m}\sum_{j=1}^{n_i} p_{ij} e^{\mu_\alpha + \frac{1}{2} + \boldsymbol{\mu}_\beta{}'\boldsymbol{x}_{ij} + \frac{1}{2}\boldsymbol{x}_{ij}{}'\boldsymbol{\Sigma}_\beta^{-1}\boldsymbol{x}_{ij}} = 1. \tag{3.42}$$

The last constraint follows by integrating both sides of the relevant constraints in (3.20) with respect to the heterogeneity distributions.

We have,

$$\hat{p}_{ij} = \frac{1}{2n_m} \frac{1}{1 + \hat{\gamma}[e^{\hat{\mu}_\alpha + \frac{1}{2}} e^{\hat{\boldsymbol{\mu}}_\beta' \boldsymbol{x}_{ij} + \frac{1}{2}\boldsymbol{x}_{ij}{}'\hat{\boldsymbol{\Sigma}}_\beta^{-1}\boldsymbol{x}_{i,j}} - 1]} \tag{3.43}$$

and

$$\hat{G}(\boldsymbol{x}) = \sum_{i=1}^{m}\sum_{j=1}^{n_i} \hat{p}_{ij} I(\boldsymbol{x}_{ij} \leq \boldsymbol{x}) \tag{3.44}$$

where $\gamma = \lambda/2n_m$ and $\lambda$ is a Lagrange multiplier, and the estimates are obtained by maximizing (3.41) after replacing $p_{ij}$ by $\hat{p}_{ij}$.

The "rest" of the $q$ distributions are now represented by the "marginalized distribution,"

$$\hat{H}(\boldsymbol{x}) = \sum_{i=1}^{m}\sum_{j=1}^{n_i} \hat{p}_{ij} e^{\hat{\mu}_\alpha + \frac{1}{2} + \hat{\boldsymbol{\mu}}_\beta' \boldsymbol{x}_{ij} + \frac{1}{2}\boldsymbol{x}_{ij}{}'\hat{\boldsymbol{\Sigma}}_\beta^{-1}\boldsymbol{x}_{ij}} I(\boldsymbol{x}_{ij} \leq \boldsymbol{x}). \tag{3.45}$$

Thus, the marginalization of the empirical likelihood results in the two distributions (3.44) and (3.45) regardless of $m$.

### Application to Tort Reform

In the United States tort reform refers to changes in the justice systems which reduced tort losses or damages. Dayaratna (2014) applied the marginalization method with a scalar density ratio

$$\frac{g_i(x)}{g(x)} = e^{\alpha_i + \beta_i x_i}; \; i = 1, \ldots, 50, \tag{3.46}$$

assuming that $\alpha_i \sim \mathrm{N}(\mu_\alpha, 1)$ and $\beta_i \sim \mathrm{N}(\mu_\beta, \sigma_\beta^2)$, to quantify the impact of tort losses resulting from medical malpractice. He used per capita tort losses from each of the 50 U.S. states in 2004 and then again in 2006, keeping per capita tort losses from 1996 as a benchmark or reference (cf. Crain et al 2009). Assuming independence, for each of the two years 2004 and 2006, the data consisted of one observation per state so that $n_1 = \cdots = n_{50} = 1$, but for 1996 all 50 state per capita were used so that $n_{51} = 50$. Probabilities of tort losses exceeding high thresholds computed from the marginalized distribution (3.45) point to a reduction of these exceedance probabilities in 2006 as compared to 2004.

# Chapter 4

# Some Asymptotic Results

*"I am incapable of conceiving infinity, and yet I do not accept finity."*
(Simone de Beauvoir, 1908-1986.)

This chapter addresses the large sample properties of the estimators of the $\alpha$'s and $\boldsymbol{\beta}$'s and of the estimator of the reference distribution function $\hat{G}(x)$ in the density ratio model (1.21) under a slight change of notation.

To align our development in this and the next chapter with many of our references particularly Qin and Zhang (1997), Zhang (2000a), Lu (2007), Wen (2013), and Katzoff et al. (2014), it is convenient to modify slightly the earlier notation and let $\boldsymbol{x}_0$ be the reference sample and $\boldsymbol{x}_1, ..., \boldsymbol{x}_m$ the associated distorted samples. Consquently, we consider the $m+1$ independent samples,

$$\begin{aligned}
\boldsymbol{x}_0 &= (x_{01}, \quad \ldots \quad, x_{0n_0})' \sim g(x) \\
\boldsymbol{x}_1 &= (x_{11}, \quad \ldots \quad, x_{1n_1})' \sim g_1(x) \\
&\qquad\qquad \vdots \\
\boldsymbol{x}_m &= (x_{m1}, \quad \ldots \quad, x_{mn_m})' \sim g_m(x),
\end{aligned} \tag{4.1}$$

and write the density ratio model as before,

$$\frac{g_j(x)}{g(x)} = \exp(\alpha_j + \boldsymbol{\beta}_j' \boldsymbol{h}(x)), \quad j = 1, \ldots, m, \tag{4.2}$$

with $p$-dimensional vectors $\boldsymbol{\beta}_j$ and normalizing scalars $\alpha_j$. Define $\alpha_0 = 0$, $\boldsymbol{\beta}_0 = \mathbf{0}$, and as was assumed all along, $\boldsymbol{h}(x)$ is a known $p$-dimensional vector valued function. Then $n = n_0 + n_1 + \cdots + n_m$, $\rho_i = n_i/n_0$, $w_j(x) = \exp(\alpha_j +$

$\boldsymbol{\beta}_j'\boldsymbol{h}(x))$, $j = 1, \ldots, m$, $w_0(x) = 1$, $\boldsymbol{\alpha} = (\alpha_1, \ldots, \alpha_m)'$, $\boldsymbol{\beta} = (\boldsymbol{\beta}_1', \ldots, \boldsymbol{\beta}_m')'$, $\boldsymbol{\theta} = (\boldsymbol{\alpha}', \boldsymbol{\beta}')'$, and

$$\boldsymbol{t} = (t_1, \ldots, t_n)' = (\boldsymbol{x}_0', \boldsymbol{x}_1', \ldots, \boldsymbol{x}_m')', \tag{4.3}$$

denotes the combined data from the $m+1$ samples. Hence, replacing $n_m$ by $n_0$ and $q$ by $m$, the log-likelihood (2.31) as a function of $\boldsymbol{\theta}$ only becomes,

$$\begin{aligned} \ell(\boldsymbol{\theta}) &= -n \log n_0 - \sum_{i=1}^{n} \log[1 + \rho_1 w_1(t_i) + \cdots + \rho_m w_m(t_i)] \\ &\quad + \sum_{j=1}^{n_1} (\alpha_1 + \boldsymbol{\beta}_1' \boldsymbol{h}(x_{1j})) + \cdots + \sum_{j=1}^{n_m} (\alpha_m + \boldsymbol{\beta}_m' \boldsymbol{h}(x_{mj})), \end{aligned} \tag{4.4}$$

and similarly replacing $1/n_m$ by $1/n_0$ and $q$ by $m$ in (2.34) we have,

$$\hat{p}_i = \frac{1}{n_0} \cdot \frac{1}{1 + \rho_1 \exp(\hat{\alpha}_1 + \hat{\boldsymbol{\beta}}_1' \boldsymbol{h}(t_i)) + \cdots + \rho_m \exp(\hat{\alpha}_m + \hat{\boldsymbol{\beta}}_m' \boldsymbol{h}(t_i))}, \tag{4.5}$$

$$\hat{G}(t) = \sum_{i=1}^{n} \hat{p}_i I(t_i \leq t), \tag{4.6}$$

and

$$\hat{G}_j(t) = \sum_{i=1}^{n} \exp(\hat{\alpha}_j + \hat{\boldsymbol{\beta}}_j' \boldsymbol{h}(t_i)) \hat{p}_i I(t_i \leq t), \quad j = 1, ..., m. \tag{4.7}$$

With

$$\nabla \equiv \left( \frac{\partial}{\partial \alpha_1}, ..., \frac{\partial}{\partial \alpha_m}, \frac{\partial}{\partial \boldsymbol{\beta}_1}, ..., \frac{\partial}{\partial \boldsymbol{\beta}_m} \right)'$$

the matrices

$$-\frac{1}{n} \nabla \nabla' \ell(\boldsymbol{\theta}) \equiv -\frac{1}{n} \boldsymbol{S}_n \to \boldsymbol{S}, \quad n \to \infty$$

and

$$\boldsymbol{\Lambda} \equiv \mathrm{Var}\left[ \frac{1}{\sqrt{n}} \nabla \ell(\boldsymbol{\theta}) \right]$$

can be derived as in Appendix 2.4. Observe that now $\boldsymbol{S}_n$ and $\boldsymbol{\Lambda}$ are $(p+1)m \times (p+1)m$ matrices.

We further observe that with $\boldsymbol{\theta}_0$ the true parameter value,

$$\mathrm{Var}(\partial\ell(\boldsymbol{\theta}_0)/\partial\boldsymbol{\theta}) = \mathrm{E}(\partial\ell(\boldsymbol{\theta}_0)/\partial\boldsymbol{\theta})(\partial\ell(\boldsymbol{\theta}_0)/\partial\boldsymbol{\theta})'$$

since

$$\mathrm{E}(\partial\ell(\boldsymbol{\theta}_0)/\partial\boldsymbol{\theta}) = \mathbf{0}.$$

To see this, consider (see (2.32))

$$\begin{aligned}
\mathrm{E}\left\{\frac{\partial\ell}{\partial\alpha_j}\right\} &= -\mathrm{E}\left\{\sum_{i=1}^{n}\frac{\rho_j w_j(t_i)}{1+\rho_k w_k(t_i)+\cdots+\rho_m w_m(t_i)}\right\}+n_j \\
&= -\sum_{u=0}^{m}\sum_{v=1}^{n_u}\mathrm{E}\left\{\frac{\rho_j w_j(x_{uv})}{\sum_{k=0}^{m}\rho_k w_k(x_{uv})}\right\}+n_j \\
&= -\sum_{u=0}^{m} n_u\int\frac{\rho_j w_j(t)}{\sum_{k=0}^{m}\rho_k w_k(t)}w_u(t)dG(t)+n_j \\
&= -n_0\rho_j\int\frac{w_j(t)}{\sum_{k=0}^{m}\rho_k w_k(t)}\sum_{u=0}^{m}\rho_u w_u(t)dG(t)+n_j \\
&= -n_0\rho_j\int w_j(t)dG(t)+n_j. \qquad (4.8)
\end{aligned}$$

Since the last integral equals 1, it follows that $\mathrm{E}(\partial\ell/\partial\alpha_j) = 0$ from the definition of $\rho_j$. Similarly,

$$\begin{aligned}
\mathrm{E}\left\{\frac{\partial\ell}{\partial\boldsymbol{\beta}_j}\right\} &= \mathrm{E}\left\{-\sum_{i=1}^{n}\frac{\rho_j w_j(t_i)\boldsymbol{h}(t_i)}{1+\rho_1 w_1(t_i)+\cdots+\rho_m w_m(t_i)}+\sum_{i=1}^{n_j}\boldsymbol{h}(x_{ji})\right\} \\
&= -\sum_{u=0}^{m}\sum_{v=1}^{n_u}\mathrm{E}\left\{\frac{\rho_j w_j(x_{uv})\boldsymbol{h}(x_{uv})}{\sum_{k=0}^{m}\rho_k w_k(x_{uv})}\right\}+\sum_{i=1}^{n_j}\mathrm{E}[\boldsymbol{h}(x_{ji})] \\
&= -\sum_{u=0}^{m} n_u\int\frac{\rho_j w_j(t)\boldsymbol{h}(t)}{\sum_{k=0}^{m}\rho_k w_k(t)}w_u(t)dG(t)+n_j\int w_j(t)\boldsymbol{h}(t)dG(t) \\
&= -\rho_j n_0\int w_j(t)\boldsymbol{h}(t)dG(t)+n_j\int w_j(t)\boldsymbol{h}(t)dG(t) = \mathbf{0} \qquad (4.9)
\end{aligned}$$

since $n_j = \rho_j n_0$.

## 4.1 Weak Convergence of $\sqrt{n}(\hat{G}(t) - G(t))$

We shall assume that the sample size ratios $\rho_j = n_j/n_0$ are positive and remain fixed as the total sample size $n = \sum_{j=0}^{m} n_j \to \infty$, and that $\boldsymbol{S}$ is positive definite.

Define the $m \times m$ diagonal matrix,

$$\boldsymbol{\rho} = diag(\rho_1, \ldots, \rho_m),$$

and the unit $m \times m$ matrix

$$\mathbf{1}_m = \begin{pmatrix} 1 & \cdots & 1 \\ \vdots & \ddots & \vdots \\ 1 & \cdots & 1 \end{pmatrix}$$

Then after some work (Qin and Zhang 1997, Zhang 2000a, Lu 2007),

$$\boldsymbol{\Sigma} \equiv \boldsymbol{S}^{-1}\boldsymbol{\Lambda}\boldsymbol{S}^{-1} = \boldsymbol{S}^{-1} - \sum_{k=0}^{m} \rho_k \begin{pmatrix} \mathbf{1}_m + \boldsymbol{\rho}^{-1} & \mathbf{0}_{m \times mp} \\ \mathbf{0}_{mp \times m} & \mathbf{0}_{mp \times mp} \end{pmatrix}. \tag{4.10}$$

Compare this to (2.19). The main steps leading to (4.10) are given in the appendix to this chapter.

We now have,

**Theorem 4.1.1** *Suppose that $\boldsymbol{S}$ is positive definite. Then,*

**(a)** *The solution $\hat{\boldsymbol{\theta}}$ of the score equations (2.32) is a strongly consistent estimator.*

**(b)** *As $n \to \infty$,*

$$\sqrt{n} \begin{pmatrix} \hat{\boldsymbol{\alpha}} - \boldsymbol{\alpha}_0 \\ \hat{\boldsymbol{\beta}} - \boldsymbol{\beta}_0 \end{pmatrix} \xrightarrow{d} \mathrm{N}_{(p+1)m}(\mathbf{0}, \boldsymbol{\Sigma}), \tag{4.11}$$

*where $\boldsymbol{\Sigma} = \boldsymbol{S}^{-1}\boldsymbol{\Lambda}\boldsymbol{S}^{-1}$.*

We shall only sketch some key points and refer to Qin and Zhang (1997) and Lu (2007) for details. To prove asymptotic normality, we expand $\partial \ell(\hat{\boldsymbol{\theta}})/\partial \boldsymbol{\theta}$ at $\boldsymbol{\theta}_0$,

$$\mathbf{0} = \frac{\partial \ell(\hat{\boldsymbol{\theta}})}{\partial \boldsymbol{\theta}} = \frac{\partial \ell(\boldsymbol{\theta}_0)}{\partial \boldsymbol{\theta}} + \frac{\partial^2 \ell(\boldsymbol{\theta}_0)}{\partial \boldsymbol{\theta}^2}\left(\hat{\boldsymbol{\theta}} - \boldsymbol{\theta}_0\right) + o_p(\delta_n),$$

where $\delta_n = |\hat{\boldsymbol{\theta}} - \boldsymbol{\theta}_0| \to \mathbf{0}$ as $n \to \infty$ since $\hat{\boldsymbol{\theta}}$ is a strongly consistent estimator of the true $\boldsymbol{\theta}_0$. Now, observe that

$$\boldsymbol{\Lambda} = \mathrm{Var}\left(\frac{1}{\sqrt{n}} \frac{\partial \ell(\boldsymbol{\theta}_0)}{\partial \boldsymbol{\theta}}\right),$$

and that $\mathrm{E}\left(\partial\ell(\boldsymbol{\theta}_0)/\partial\boldsymbol{\theta}\right) = \mathbf{0}$. Then from the central limit theorem

$$\frac{1}{\sqrt{n}}\frac{\partial\ell(\boldsymbol{\theta}_0)}{\partial\boldsymbol{\theta}} \xrightarrow{d} N_{(m+1)p}(\mathbf{0}, \boldsymbol{\Lambda}). \tag{4.12}$$

The fact that $-(1/n)\boldsymbol{S}_n \to \boldsymbol{S}$ together with Slutsky's theorem gives,

$$\begin{aligned}\sqrt{n}\left(\hat{\boldsymbol{\theta}} - \boldsymbol{\theta}_0\right) &= \boldsymbol{S}^{-1}\frac{1}{\sqrt{n}}\frac{\partial\ell(\boldsymbol{\theta}_0)}{\partial\boldsymbol{\theta}} + o_p(1) \\ &\xrightarrow{d} \mathrm{N}_{(p+1)m}(\mathbf{0}, \boldsymbol{S}^{-1}\boldsymbol{\Lambda}\boldsymbol{S}^{-1}).\end{aligned}$$

Thus, from (4.10), $\sqrt{n}\left(\hat{\boldsymbol{\theta}} - \boldsymbol{\theta}_0\right) \xrightarrow{d} \mathrm{N}_{(p+1)m}(\mathbf{0}, \boldsymbol{\Sigma})$, and hence the asymptotic distribution of $\sqrt{n}(\hat{\boldsymbol{\beta}} - \boldsymbol{\beta}_0)$ is normal with mean $\mathbf{0}$ and with a covariance matrix obtained from the appropriate part in $\boldsymbol{\Sigma}$. We have used this fact to get the asymptotic distribution of the $\mathcal{X}_1$ test statistic in (2.37) when testing the equidistribution hypothesis $\mathrm{H}_0 : \boldsymbol{\beta}_0 = \mathbf{0}$.

The asymptotic behavior of $\hat{G}(x)$ has been studied by Vardi (1982), Qin and Zhang (1997), Gilbert (2000), Zhang (2000a), and Lu (2007). In particular, $\sqrt{n}(\hat{G}(t) - G(t))$ converges to a Gaussian process. To state this precisely, define for $j = 1, ..., m$

$$A_j(t) = \int \frac{w_j(y)I(y \le t)}{\sum_{k=1}^m \rho_k w_k(y)} dG(y), \quad \boldsymbol{B}_j(t) = \int \frac{w_j(y)\boldsymbol{h}(y)I(y \le t)}{\sum_{k=1}^m \rho_k w_k(y)} dG(y)$$

and let

$$\bar{\boldsymbol{A}}(t) = (A_1(t), \ldots, A_m(t))', \quad \bar{\boldsymbol{B}}(t) = (\boldsymbol{B}_1'(t), \ldots, \boldsymbol{B}_m'(t))',$$

and recall $\boldsymbol{\rho} = \mathrm{diag}\{\rho_1, \ldots, \rho_m\}$ is an $m \times m$ matrix. We also need the empirical distribution $\tilde{G}$ from the reference sample, $\boldsymbol{x}_0 = (x_{01}, \ldots, x_{0n_0})$,

$$\tilde{G}(t) = \frac{1}{n_0}\sum_{i=1}^{n_0} I_{[x_{0i} \le t]}. \tag{4.13}$$

To study the asymptotic distribution of $\sqrt{n}(\hat{G} - G)$ we first express $\sqrt{n}(\hat{G}(t) - G(t))$ as a sum of two components

$$\sqrt{n}(\hat{G}(t) - \tilde{G}(t)) + \sqrt{n}(\tilde{G}(t) - G(t)).$$

Since the asymptotic properties of $\sqrt{n}(\tilde{G}(t) - G(t))$ are well known from the theory of empirical processes, a key step is to prove the weak convergence of $\sqrt{n}(\hat{G}(t) - \tilde{G}(t))$ to a Gaussian process. In the two-sample case this has been studied in Qin and Zhang (1997). The multiple sample generalization is given in Lu (2007) as follows, assuming that all moments with respect to the reference distribution are finite.

**Theorem 4.1.2** *The process $\sqrt{n}(\hat{G} - \tilde{G})$ converges weakly to a zero-mean Gaussian process $W$ with continuous sample paths in $D[-\infty, \infty]$, and the covariance matrix is determined by*

$$E[W(t)W(s)] = \sum_{k=0}^{m} \rho_k \sum_{j=1}^{m} \rho_j A_j(t \wedge s) - \left(\bar{\boldsymbol{A}}'(t)\boldsymbol{\rho}, \bar{\boldsymbol{B}}'(t)(\boldsymbol{\rho} \otimes I_p)\right) \boldsymbol{S}^{-1} \begin{pmatrix} \boldsymbol{\rho}\bar{\boldsymbol{A}}(s) \\ (\boldsymbol{\rho} \otimes I_p)\bar{\boldsymbol{B}}(s) \end{pmatrix}.$$

Then we have from Lu (2007).

**Theorem 4.1.3** *The process $\sqrt{n}(\hat{G}(t) - G(t))$ converges weakly to a zero-mean Gaussian process in $D[-\infty, \infty]$, with covariance matrix given by*

$$\begin{aligned} &\mathrm{Cov}\{\sqrt{n}(\hat{G}(t) - G(t)), \sqrt{n}(\hat{G}(s) - G(s))\} = \\ &\sum_{k=1}^{m} \rho_k \left( G(t \wedge s) - G(t)G(s) - \sum_{j=1}^{q} \rho_j A_j(t \wedge s) \right) \\ &+ \left(\bar{\boldsymbol{A}}'(s)\boldsymbol{\rho}, \bar{\boldsymbol{B}}'(s)(\boldsymbol{\rho} \otimes \boldsymbol{I}_p)\right) \boldsymbol{S}^{-1} \begin{pmatrix} \boldsymbol{\rho}\bar{\boldsymbol{A}}(t) \\ (\boldsymbol{\rho} \otimes \boldsymbol{I}_p)\bar{\boldsymbol{B}}(t) \end{pmatrix} \end{aligned} \tag{4.14}$$

where $D[-\infty, \infty]$ is the space of all right continuous functions on $[-\infty, \infty]$ with left limits. This result reduces to that of Zhang (2000a) who proved the same fact for $m = 1$. We provide the main steps of the proof in Appendix 4.3. For a complete proof of Theorem 4.1.3 the reader is referred to Lu (2007). The immediate application of Theorem 4.1.3 is in the construction of pointwise symmetric confidence intervals for $G(t)$ for any given $t$ which we shall apply in the next chapter in the interval estimation of small tail probabilities.

## 4.2 Goodness of Fit

We observe that the empirical distribution estimator $\tilde{G}$ from the reference sample $\boldsymbol{x}_0$ only is a model free estimator of the reference $G$, while the semi-parametric estimator $\hat{G}$ is derived under the density ratio model. Hence, model validation could be judged to a reasonable extent by the closeness of these two estimators. Clearly, relatively large discrepancy or difference between $\hat{G}$ and $\tilde{G}$ calls into question the appropriateness as well as the usefulness of the model. We have used this idea graphically in Figure 3.1 in

Section 3.4.1 by plotting $\tilde{G}$ versus $\hat{G}$, and earlier in formulating the goodness of fit statistic in (2.22),

$$\Delta_n = \sup_{-\infty \le t \le \infty} \sqrt{n}\, |\hat{G}(t) - \tilde{G}(t)|, \tag{4.15}$$

which was used in the radar application in Section 2.2.7. Recall that in that application the distribution of $\Delta_n$ was approximated from numerous different reflectivity samples. In general, however, we do not have access to such large amounts of data, and the distribution of $\Delta_n$ must be approximated by other computational means. And, all the more so since getting the distribution of $\Delta_n$ analytically is problematic.

The Kolmogorov-Smirnov type statistic $\Delta_n$ (4.15) is based on the distance between $\hat{G}$ and $\tilde{G}$. Clearly, we could have used any of the $\hat{G}_j$ and the corresponding empirical distribution from the $j$th sample $\tilde{G}_j$, or even a combination of these estimated distributions in the formulation of similar goodness of fit statistics. It seems that (4.15) is a "natural" choice and we shall adhere to it.

Let $\hat{G}, \hat{G}_1, \ldots, \hat{G}_m$ be the estimated distribution functions from the combined sample $(\boldsymbol{x}_0, \boldsymbol{x}_1, \ldots, \boldsymbol{x}_m)$ as given in (4.6) and (4.7), and let

$$\boldsymbol{x}_0^*, \boldsymbol{x}_1^*, \ldots, \boldsymbol{x}_m^*$$

be $m+1$ "bootstrap" random samples, with sizes $n_0, n_1, ..., n_m$, generated from $\hat{G}$, $\hat{G}_1$, ..., $\hat{G}_m$, respectively. We now repeat all the calculations above using the fused bootstrap data $(\boldsymbol{x}_0^*, \boldsymbol{x}_1^*, \ldots, \boldsymbol{x}_m^*)$. In this way we get the estimates $(\hat{\boldsymbol{\alpha}}^*, \hat{\boldsymbol{\beta}}^*)$ and the estimated reference $\hat{G}^*$ from $(\boldsymbol{x}_0^*, \boldsymbol{x}_1^*, \ldots, \boldsymbol{x}_m^*)$. Let $\tilde{G}^*$ be the empirical distribution from $\boldsymbol{x}_0^*$ only. Then the analog of the test statistic $\Delta_n$ is given by

$$\Delta_n^* = \sup_{-\infty \le t \le \infty} \sqrt{n}\, |\hat{G}^*(t) - \tilde{G}^*(t)|.$$

It can be shown that $\Delta_n^*$ has the same limiting distribution as that of $\Delta_n$ under the density ratio model (4.2) (Qin and Zhang 1997, Zhang 2000a). Consequently, we can approximate the quantiles of $\Delta_n$ by those of $\Delta_n^*$, where the distribution of $\Delta_n^*$ is estimated from many bootstrap samples $(\boldsymbol{x}_0^*, \boldsymbol{x}_1^*, \ldots, \boldsymbol{x}_m^*)$. Thus, the validity of model (4.2) is rejected for large values of $\Delta_n$ as judged by the $p$-values (really their estimates)

$$P(\Delta_n^* > \Delta_{n,obs})$$

where $\Delta_{n,obs}$ is the observed value of $\Delta_n$. Relatively small $p$-values challenge the probity or goodness of the density ratio model.

Building on an earlier work of White (1982) on maximum likelihood estimation of misspecified models, Zhang (2001) proposed a Wald-type quadratic form, which he calls "information matrix statistic," for testing the goodness of fit of the logistic regression model in case-control problems as an alternative to the bootstrap method of Qin and Zhang (1997). Bondell (2007) discussed a goodness-of-fit test for the logistic regression model under case-control sampling measuring the discrepancy between $\hat{G}$ and $\tilde{G}$ by the integrated squared error resulting from the corresponding kernel density estimators. Rewriting (3.28) and (3.29) for a given kernel $K$ with a fixed bandwidth as

$$\begin{aligned}\hat{f}(x) &= \int K(x-y)d\hat{G}(y)\\ \tilde{f}(x) &= \int K(x-y)d\tilde{G}(y),\end{aligned}$$

the integrated squared error is defined by the integral of the squared difference,

$$I_n \equiv n \int \left\{\hat{f}(x) - \tilde{f}(x)\right\}^2 dx. \tag{4.16}$$

Under the hypothesis that $\boldsymbol{x}_0 \sim f_0(x)$, $\boldsymbol{x}_1 \sim \exp(\alpha + x\beta)f_0(x)$ (here $p = 1$) the distribution of $I_n$ involves an intractable infinite linear combination of chi-square random variables. Hence, the author advocates a bootstrap procedure as in Qin and Zhag (1997) to derive an approximate distribution of the test statistic $I_n$. It seems that, in goodness-of-fit vis-à-vis logistic regression, switching to the kernel density compares favorably power-wise with the methods of Qin and Zhang (1997) and of Zhang (2001). Testing the goodness of fit of the general density ratio model using the statistic (4.16) is discussed in Cheng and Chu (2004) who provide the asymptotic distribution of this statistic.

Another possible alternative is the $R^2_{\alpha,k}$ statistic discussed in Voulgaraki et al. (2012). Consider the $i$th sample of size $n_i$, and let $x_\alpha$ be the number of times $\hat{G}_i$ falls in the estimated $1-\alpha$ confidence interval obtained from the corresponding empirical $\tilde{G}_i$, both evaluated at the sample points. Define

$$R^2_{\alpha,k} = 1 - \exp\left\{-\left(\frac{x_\alpha}{n_i - x_\alpha}\right)^k\right\} \tag{4.17}$$

where $k > 0$, and $k$ and $\alpha$ are parameters chosen by the user. Observe that $R^2_{\alpha,k}$ takes values between 0 and 1, being close to 1 when $x_\alpha$ approaches

$n_i$ and close to 0 when $x_\alpha$ is close to 0. This statistic has been used in connection with the TGCT example in Section 3.4.1 in Voulgaraki et al. (2012), assuming high values echoing the diagnostic plots in Figure 3.1, and in public policy problems in Dayaratna (2014).

### 4.2.1 Examples of Goodness of Fit

Pairs of samples $\boldsymbol{x}_0, \boldsymbol{x}_1$, $n_0 = n_1 = 100$, were generated from various distributions, giving the corresponding $\hat{G}, \hat{G}_1$ assuming $\boldsymbol{h}(x) = (x, x^2)'$. Then, maintaining the same sample sizes of 100 throughout, 500 bootstrap samples $\boldsymbol{x}_0^*, \boldsymbol{x}_1^*$ were generated from $\hat{G}, \hat{G}_1$. Table 4.1 gives the estimated $p$-values based on 500 replications of $\Delta_n^*$ in each case. We see that when the reference distribution is N(0,1), normal or close to normal "case" samples give large p-values, whereas case samples from distributions far from normal give small $p$-values. The relatively large $p$-values in the two uniform cases will be explained in the next chapter. More precisely, the table shows that when the reference distribution is N(0,1) the density ratio model with $\boldsymbol{h}(x) = (x, x^2)'$ is rejected for $g_1$ from $t_{(1)}$, $t_{(2)}$, Weibull(1.2,1), and Cauchy(0,1) (another run of $t_{(1)}$), but is plausible in the other cases in Table 4.1.

Table 4.1: $p$-values $P(\Delta_n^* > \Delta_{n,obs})$ from 500 replications of $\Delta_n^*$.

| $\boldsymbol{x}_0$ | $\boldsymbol{x}_1$ | $\Delta_{n,obs}$ | $p$-value |
|---|---|---|---|
| N(0,1) | $t_{(1)}$ | 2.401537 | 0.0000 |
| N(0,1) | $t_{(2)}$ | 0.801353 | 0.0440 |
| N(0,1) | $t_{(15)}$ | 0.528735 | 0.4460 |
| N(0,1) | Unif(-3,4) | 0.411866 | 0.8020 |
| N(0,1) | Unif(-5,5) | 0.354214 | 0.9460 |
| N(0,1) | N(3,25) | 0.343834 | 0.9440 |
| N(0,1) | Weibull(1.2,1) | 0.977296 | 0.0020 |
| N(0,1) | Cauchy(0,1) | 2.854246 | 0.0000 |

Consider another example where $\boldsymbol{x}_0 \sim N(0,1)$, and $\boldsymbol{x}_1 \sim N(0,4)$, $n_0 = n_1 = 100$. So, the correct tilt function is $h(x) = x^2$. Running the bootstrap method (with 500 replications) we get $\Delta_{n,obs} = 0.4640$ and $P(\Delta_n^* > 0.4640) = 0.62$, lending support to the model. But by using the incorrect $h(x) = x$ we obtain a misspecified density ratio model where now, with the same $\boldsymbol{x}_0, \boldsymbol{x}_1$, the discrepancy measure is $\Delta_{n,obs} = 2.5191$ and $P(\Delta_n^* > 2.5191) = 0.0000$, which invalidates the model. Repeating this experiment many times gives very similar results where the $p$-values are high for a correctly specified tilt function, and very low otherwise. This is yet another example where goodness-of-fit sheds light on the choice of the tilt function, as was already commented on in Section 2.2.4.

## 4.3 Appendix

### 4.3.1 Representation of $\boldsymbol{\Sigma}$

A complete derivation of $\boldsymbol{\Sigma}$ in (4.10) is given in Lu (2007). Some key steps are as follows.

We first define,

$$\begin{aligned}
A_{jj'} &= \int \frac{w_j(t)w_{j'}(t)}{\sum_{k=0}^{m}\rho_k w_k(t)} dG(t) \\
\boldsymbol{B}_{jj'} &= \int \frac{w_j(t)w_{j'}(t)\boldsymbol{h}(t)}{\sum_{k=0}^{m}\rho_k w_k(t)} dG(t) \\
\boldsymbol{C}_{jj'} &= \int \frac{w_j(t)w_{j'}(t)\boldsymbol{h}(t)\boldsymbol{h}'(t)}{\sum_{k=0}^{m}\rho_k w_k(t)} dG(t) \\
\boldsymbol{E}_j &= \mathrm{E}(\boldsymbol{h}(x_{ji})) = \int w_j(t)\boldsymbol{h}(t)dG(t) \quad \bar{E}_j = \int w_j(t)\boldsymbol{h}(t)\boldsymbol{h}'(t)dG(t)
\end{aligned}$$

where $\boldsymbol{B}_{jj'}$ and $\boldsymbol{E}_j$ are $p \times 1$ vectors, and $\boldsymbol{C}_{jj'}, \bar{\boldsymbol{E}}_j$ and are $p \times p$ matrices.

Further define the matrices,

$$\boldsymbol{A} = (A_{ij})_{m\times m}, \ \boldsymbol{B} = (\boldsymbol{B}_{ij})_{mp\times m}, \ \boldsymbol{C} = (\boldsymbol{C}_{ij})_{mp\times mp},$$

$$\boldsymbol{E} = \begin{pmatrix} \boldsymbol{E}_1 & \cdots & \tilde{\mathbf{0}} \\ \vdots & \ddots & \vdots \\ \tilde{\mathbf{0}} & \cdots & \boldsymbol{E}_m \end{pmatrix}_{mp\times m} \quad \bar{\boldsymbol{E}} = \begin{pmatrix} \bar{\boldsymbol{E}}_1 & \cdots & \hat{\mathbf{0}} \\ \vdots & \ddots & \vdots \\ \hat{\mathbf{0}} & \cdots & \bar{\boldsymbol{E}}_m \end{pmatrix}_{mp\times mp} \tag{4.18}$$

where $\tilde{\mathbf{0}}$ is a $p \times 1$ vector of 0's, and $\hat{\mathbf{0}}$ is a $p \times p$ matrix of 0's.

We now partition $\boldsymbol{\Lambda}$ and $\boldsymbol{S}$,

$$\boldsymbol{\Lambda} = \mathrm{Var}\left(\frac{1}{\sqrt{n}}\frac{\partial \ell}{\partial \boldsymbol{\theta}}\right) = \frac{1}{\sum_{k=0}^{m}\rho_k}\begin{pmatrix} \Lambda_{11} & \Lambda_{12} \\ \Lambda_{21} & \Lambda_{22} \end{pmatrix}, \tag{4.19}$$

and

$$\boldsymbol{S} = \frac{1}{\sum_{k=0}^{m}\rho_k}\begin{pmatrix} S_{11} & S_{12} \\ S_{21} & S_{22} \end{pmatrix}, \tag{4.20}$$

where

$$
\begin{aligned}
S_{11} &= \boldsymbol{\rho} - \boldsymbol{\rho A \rho} \\
S_{12} &= \boldsymbol{\rho E}' - \boldsymbol{\rho B}'(\boldsymbol{\rho} \otimes \boldsymbol{I}_p) \\
S_{21} &= S_{12}' = \boldsymbol{E\rho} - (\boldsymbol{\rho} \otimes \boldsymbol{I}_p)\boldsymbol{B\rho} \\
S_{22} &= (\boldsymbol{\rho} \otimes \boldsymbol{I}_p)\bar{\boldsymbol{E}} - (\boldsymbol{\rho} \otimes \boldsymbol{I}_p)\boldsymbol{C}(\boldsymbol{\rho} \otimes \boldsymbol{I}_p).
\end{aligned} \tag{4.21}
$$

and

$$
\begin{aligned}
\Lambda_{11} &= \boldsymbol{\rho A \rho} - \boldsymbol{\rho A \rho A \rho} - \boldsymbol{\rho 1}_m \boldsymbol{\rho} + \boldsymbol{\rho A \rho 1}_m \boldsymbol{\rho} + \boldsymbol{\rho 1}_m \boldsymbol{\rho A \rho} \\
&\quad - \boldsymbol{\rho A \rho 1}_m \boldsymbol{\rho A \rho} \\
\Lambda_{12} &= \boldsymbol{\rho A E}'(\boldsymbol{\rho} \otimes \boldsymbol{I}_p) - \boldsymbol{\rho A \rho B}'(\boldsymbol{\rho} \otimes \boldsymbol{I}_p) - \boldsymbol{\rho 1}_m \boldsymbol{E}'(\boldsymbol{\rho} \otimes \boldsymbol{I}_p) \\
&\quad + \boldsymbol{\rho A \rho 1}_m \boldsymbol{E}'(\boldsymbol{\rho} \otimes \boldsymbol{I}_p) + \boldsymbol{\rho 1}_m \boldsymbol{\rho B}'(\boldsymbol{\rho} \otimes \boldsymbol{I}_p) \\
&\quad - \boldsymbol{\rho A \rho 1}_m \boldsymbol{B}'(\boldsymbol{\rho} \otimes \boldsymbol{I}_p) \\
\Lambda_{21} &= \Lambda_{12}' = (\boldsymbol{\rho} \otimes \boldsymbol{I}_p)\boldsymbol{E A \rho} - (\boldsymbol{\rho} \otimes \boldsymbol{I}_p)\boldsymbol{B \rho A \rho} - (\boldsymbol{\rho} \otimes \boldsymbol{I}_p)\boldsymbol{E 1}_m \boldsymbol{\rho} \\
&\quad + (\boldsymbol{\rho} \otimes \boldsymbol{I}_p)\boldsymbol{E 1}_m \boldsymbol{\rho A \rho} + (\boldsymbol{\rho} \otimes \boldsymbol{I}_p)\boldsymbol{B \rho 1}_m \boldsymbol{\rho} \\
&\quad - (\boldsymbol{\rho} \otimes \boldsymbol{I}_p)\boldsymbol{B 1}_m \boldsymbol{\rho A \rho} \\
\Lambda_{22} &= -(\boldsymbol{\rho} \otimes \boldsymbol{I}_p)\boldsymbol{C}(\boldsymbol{\rho} \otimes \boldsymbol{I}_p) - (\boldsymbol{\rho} \otimes \boldsymbol{I}_p)\boldsymbol{B \rho B}'(\boldsymbol{\rho} \otimes \boldsymbol{I}_p) \\
&\quad + (\boldsymbol{\rho} \otimes \boldsymbol{I}_p)\boldsymbol{B E}'(\boldsymbol{\rho} \otimes \boldsymbol{I}_p) + (\boldsymbol{\rho} \otimes \boldsymbol{I}_p)\boldsymbol{E B}'(\boldsymbol{\rho} \otimes \boldsymbol{I}_p) + (\boldsymbol{\rho} \otimes \boldsymbol{I}_p)\boldsymbol{V} \\
&\quad - (\boldsymbol{\rho} \otimes \boldsymbol{I}_p)\boldsymbol{E 1}_m \boldsymbol{E}'(\boldsymbol{\rho} \otimes \boldsymbol{I}_p) + (\boldsymbol{\rho} \otimes \boldsymbol{I}_p)\boldsymbol{B \rho 1}_m \boldsymbol{E}'(\boldsymbol{\rho} \otimes \boldsymbol{I}_p) \\
&\quad + (\boldsymbol{\rho} \otimes \boldsymbol{I}_p)\boldsymbol{E 1}_m \boldsymbol{\rho B}'(\boldsymbol{\rho} \otimes \boldsymbol{I}_p) - (\boldsymbol{\rho} \otimes \boldsymbol{I}_p)\boldsymbol{B \rho 1}_m \boldsymbol{\rho B}'(\boldsymbol{\rho} \otimes \boldsymbol{I}_p),
\end{aligned} \tag{4.22}
$$

where $\boldsymbol{I}_p$ is the $p \times p$ identity matrix, and $\otimes$ denotes kronecker product. Then $\boldsymbol{\Lambda}$ and $\boldsymbol{S}$ are related:

$$
\boldsymbol{\Lambda} = \boldsymbol{S} - \frac{1}{\sum_{k=0}^{m} \rho_k} \begin{pmatrix} S_{11} \\ S_{21} \end{pmatrix} (\boldsymbol{1}_m + \boldsymbol{\rho}^{-1}) \begin{pmatrix} S_{11} & S_{12} \end{pmatrix}.
$$

Therefore, with obvious notation,

$$
\begin{aligned}
\boldsymbol{\Sigma} &= \boldsymbol{S}^{-1}\boldsymbol{\Lambda}\boldsymbol{S}^{-1} \\
&= \boldsymbol{S}^{-1} - \sum_{k=0}^{m} \rho_k \begin{pmatrix} S_{11} & S_{12} \\ S_{21} & S_{22} \end{pmatrix}^{-1} \begin{pmatrix} S_{11} \\ S_{21} \end{pmatrix} \\
&\times \ (\mathbf{1}_m + \boldsymbol{\rho}^{-1}) \begin{pmatrix} S_{11} & S_{12} \end{pmatrix} \begin{pmatrix} S_{11} & S_{12} \\ S_{21} & S_{22} \end{pmatrix}^{-1} \\
&= \boldsymbol{S}^{-1} - \sum_{k=0}^{m} \rho_k \begin{pmatrix} 1 & 0 \\ 0 & 0 \end{pmatrix} (\mathbf{1}_m + \boldsymbol{\rho}^{-1}) \begin{pmatrix} 1 & 0 \\ 0 & 0 \end{pmatrix} \\
&= \boldsymbol{S}^{-1} - \sum_{k=0}^{m} \rho_k \begin{pmatrix} \mathbf{1}_m + \boldsymbol{\rho}^{-1} & 0 \\ 0 & 0 \end{pmatrix}.
\end{aligned}
$$

which proves (4.10).

### 4.3.2 Proof of Theorem 4.1.3

Regarding Theorem 4.1.3, the complete proof is quite long and is given in Lu (2007), and for $m = 1$ also in Zhang (2000a). We shall only provide the main steps of the proof. First we write

$$\sqrt{n}(\hat{G}(t) - G(t)) = \sqrt{n}(\hat{G}(t) - \tilde{G}(t)) + \sqrt{n}(\tilde{G}(t) - G(t))$$

where

$$\tilde{G}(t) = \frac{1}{n_0} \sum_{i=1}^{n_0} I_{[x_{0i} \le t]}. \tag{4.23}$$

is the empirical distribution of the reference sample $\boldsymbol{x}_0$ only. As the asymptotic properties of $\sqrt{n}(\tilde{G}(t) - G(t))$ are well-known, the objective is to prove the weak convergence of $\sqrt{n}(\hat{G}(t) - \tilde{G}(t))$. By the strong consistency of $\hat{\boldsymbol{\theta}}$ and a Taylor expansion of $\hat{G}(t)$ at the true parameter $\boldsymbol{\theta}_0$, $\hat{G}(t)$ is approximated uniformly in $t$ by $H_1(t) - H_2(t)$ where $H_1(t) \equiv H_1(t; \boldsymbol{\alpha}_0, \boldsymbol{\beta}_0)$ is defined by

$$H_1(t; \boldsymbol{\alpha}, \boldsymbol{\beta}) = \frac{1}{n_0} \cdot \sum_{i=1}^{n} \frac{I(t_i \le t)}{\sum_{k=0}^{m} \rho_k w_k(t_i; \alpha_k, \beta_k)},$$

and

$$H_2(t) = \frac{1}{n}\Big(\bar{\boldsymbol{A}}'(t)\boldsymbol{\rho}, \bar{\boldsymbol{B}}'(t)(\boldsymbol{\rho} \otimes I_p)\Big)\boldsymbol{S}^{-1} \begin{pmatrix} \frac{\partial \ell(\boldsymbol{\alpha}_0, \boldsymbol{\beta}_0)}{\partial \boldsymbol{\alpha}} \\ \frac{\partial \ell(\boldsymbol{\alpha}_0, \boldsymbol{\beta}_0)}{\partial \boldsymbol{\beta}} \end{pmatrix}. \tag{4.24}$$

Hence the asymptotic behavior of $\sqrt{n}(\hat{G}(t) - \tilde{G}(t))$ is equivalent to that of $\sqrt{n}(H_1(t) - H_2(t) - \tilde{G}(t))$, which involves only the true parameter $\boldsymbol{\theta}_0$. The weak convergence of the finite-dimensional distributions of $\sqrt{n}(H_1(t) - H_2(t) - \tilde{G}(t))$ follows from the multivariate central limit theorem after obtaining the variance-covariance structure. Tightness of $\sqrt{n}(H_1(t) - H_2(t) - \tilde{G}(t))$ is shown by noting that both $\sqrt{n}(H_1(t) - \tilde{G}(t))$ and $\sqrt{n}H_2(t)$ can be decomposed into sums of empirical processes.

# Chapter 5

# Out of Sample Fusion

*"I want to enter the outside."*
(A little boy, 2009.)

*"Colombian literary titan fused myth and reality."*
(Gabriel Garcia Marquez, 1927-2014. The Washington Post, April 18, 2014.)

## 5.1 Introduction

In the earlier discussion in the first few chapters, independent samples were combined or fused by assuming the density ratio model. We have seen that the implementation of the model requires at least two samples, of which one originates from a reference distribution. Suppose, however, we have only a single real data sample whose parent distribution is of interest, then there is no reason why the fusion samples cannot be computer generated "fake" data.

It is convenient to think of the given real sample as a reference sample $\boldsymbol{x}_0$, and denote the artificial data by $\boldsymbol{x}_1$.

There are several advantages to fusing real and artificial data, chiefly among them is the fact that what is generated can be controlled to make sure the density ratio model holds to a reasonable degree, the implication being that although what is generated is indeed artificial, nonetheless it is not arbitrary. Thus, when the reference distribution is symmetric we might fuse the given reference sample $\boldsymbol{x}_0$ with artificial symmetric data, and count reference data might be fused with artificial count data, and so on, and all along goodness of fit criteria can be used to test the validity of the density ratio model. Second, we can control the number of artificial samples and their sizes, and third, we can *repeat* the integration of real and artificial data many times to, for example, check the reliability of the results, and more generally obtain improved inference about the reference distribution. Fourth, since more data (albeit artificial) are used in addition to $\boldsymbol{x}_0$, the method produces on average shorter confidence intervals for small threshold or survival probabilities $1 - G(T)$ for relatively large thresholds $T$ as compared with methods which use the reference sample only. The last point underscores cases where the underlying sample is relatively small but can be augmented by artificial data to produce more precise inference under some conditions.

Since *external* data are fused with real data samples, we refer to this as *out of sample fusion* (OSF) (Zhou 2013, Katzoff et al. 2014). In the spirit of importance sampling, Fokianos and Qin (2008) use this idea in the problem of estimating the normalizing constant of a parametric probability distribution, generating a suitable sample in addition to a given one to obtain a density ratio model with a non-linear tilt as in (1.12) in Section 1.1. Similarly, Fithian and Wager (2015) study the estimation of the tail of heavy-tailed distribution given a relatively small sample, and a much larger background sample from another distribution, assuming that the tails of the two distributions are connected by an exponential tilt function motivated from extreme value theory.

A variation of OSF is *repeated out of sample fusion* (ROSF) where a given sample is fused repeatedly with many computer generated samples. Unlike the bootstrap method which is a *within-sample* procedure, ROSF is a *beyond-sample* procedure based on externally generated data: *thinking out of the sample* as it were. In direct analogy with mirror images, we might think of each fusion of a given sample with artificial or "fake" data as a different mirror image taken from a different angle, where each image sheds additional light on the object of interest. Here the object of interest is the population from which the given sample was drawn. Under certain conditions, linking ROSF to the behavior of the maximum of independent and identically distributed (i.i.d.) observations is useful in interval estimation of small tail probabilities (Kedem, et al. 2016).

Now, suppose we buy into the out of sample fusion idea, what type of artificial data is appropriate? To answer this, assume that a reference $g$ and the tilted $g_1$ satisfy the density ratio model for some $\boldsymbol{h}$. For simplicity assume that $g$ and $g_1$ are supported over $(0, \infty)$. If the densities are right-truncated at the same point $b > 0$, then the density ratio still holds with the same $\boldsymbol{h}$, and the same is true if $g_1$ is replaced by a uniform distribution over $(0, b)$. More generally, if $g$ is supported, say, over $(0, \infty)$, and

$g_1$ is a uniform density with sufficiently large support, then in many cases the density ratio is satisfied approximately for some $\boldsymbol{h}$. Thus, fusion with uniformly generated data is a possibility. And in fact a most important one. We shall demonstrate that this "works" surprisingly well in the prediction of future exceedance probabilities and in interval estimation of very small tail or survival probabilities.

## 5.2 Normal or Not?

To motivate the idea of repeated out of sample fusion (ROSF) we consider first very briefly the problem of testing the normality of a given sample. Such a problem arises, for example, in linear regression where it is of interest to test whether the residuals follow the normal distribution in order to lend support to the assumption of normal observations. We do not wish to provide a comprehensive treatment of normality tests. The goal is to point to the applicability as well as usefulness of an additional, very different though, tool in testing normality by repeated out of sample fusion.

The idea is to fuse a given sample repeatedly with computer generated normal samples and perform multiple equidistribution two-sample tests discussed in Sections 2.2.6 and 2.3.1. Since the generated data are controlled, we can try generating samples which match the given sample reasonably well. When an acceptable matching takes place the $p$-values increase, and we may identify the given sample with its *known* generated counterparts. As an overall matching measure we can use, for example, the average $p$-values. Certainly, various modifications of this are possible.

Taking the given sample as the reference $\boldsymbol{x}_0$ (fixed) and the generated samples as $\boldsymbol{x}_1$, the following repeated application of function SP2XXSQK() from Section 2.4.3 performs the desired matching, supplying the corresponding average $p$-values from 500 repeated fusions of the given *fixed* $\boldsymbol{x}_0$ and the *changing* generated $\boldsymbol{x}_1$, assuming $\boldsymbol{h} = (x, x^2)$. Here $\boldsymbol{x}_0 \sim \mathrm{N}(0,1)$, $n_0 = 100$, and $\boldsymbol{x}_1 \sim \mathrm{N}(\mathrm{mean}, \mathrm{se}^2)$, $n_1 = 1000$.

```
pval_LR <- 0; pval_chi1 <- 0
for(i in 1:500){
x1 <- rnorm(1000,mean,se)
pval_chi1[i] <- SP2XXSQK(x1,x0,0.05,.5)$chi1
pval_LR[i] <- SP2XXSQK(x1,x0,0.05,.5)$pval_LR}
mean(pval_LR)
mean(pval_chi1)
```

For (mean,se)=(0.2,1.1) the mean $p$-values of $p-LR, p-\mathcal{X}_1$ are 0.18580, and 0.18567, respectively, which increase substantially for (mean,se)=(-0.01,1.1) giving 0.78206, and 0.77509, respectively, and reaching 0.90452, and 0.90571, respectively for (mean,se)=(-0.01295905,1.107243). In this example, taking the matching samples as logistic or $t$ gives average $p$-values close to 0.

## 5.3 TS Prediction by Out of Sample Fusion

The problem of time series (TS) prediction is revisited in this section, given a time series that is regressed linearly or non-linearly on covariate time records. The time series may be stationary or nonstationary, and it may be relatively short. The goal is to estimate the *predictive distribution* conditional on past covariates by fusing the residuals with computer generated data. The ideas here are very different from Bayesian methods where the predictive distribution is obtained by a clever integration of the posterior distribution (De Oliveira et al. 1997).

The method presented hinges on well behaved residuals, assumed equidistributed, and the fact that fusion with artificial data means larger combined samples and hence more precise inference under the density ratio model. Since in many cases the residuals of regression models "look normal", fusion with computer generated normal samples is sensible, however, we do not assume that the residuals follow a normal distribution.

Consider the following time series regression model,

$$x_{t+1} = f(\boldsymbol{z}_t) + \epsilon_{t+1}, \quad t = 1, 2, ..., n_0 \tag{5.1}$$

where the vector $\boldsymbol{z}_t$ contains past values of covariate time series possibly including even past values of the response $x_t$, and where $\epsilon_t$ is an independent noise component. As values from $x_t$ can be highly dependent and/or not necessarily equidistributed, *we approach time series prediction through the distribution of the noise component estimated under a density ratio assumption* (Kedem et al. 2005,2008, Kedem and Gagnon 2010).

Assume that $\epsilon_t \sim G$ for every $t$. Then, if an additional source of data (real or artificial) $\eta_t$, $t = 1, 2, ...n_1$, is available, we can fuse the $\epsilon$'s and $\eta$'s to get an estimate $\hat{G}$ under a density ratio model for some tilt function $\boldsymbol{h}$. Denote the combined "data" of size $n = n_0 + n_1$ by

$$\boldsymbol{\tau} = (\tau_1, ..., \tau_n) \equiv \{(\epsilon_1, ..., \epsilon_{n_0}), (\eta_1, ..., \eta_{n_1})\}. \tag{5.2}$$

Since $x_{t+1} = f(\boldsymbol{z}_t) + \epsilon_{t+1}$ and $\epsilon_{t+1} \sim G$, we obtain the following approximation of the *predictive distribution* at $t+1$ conditional on $\boldsymbol{z}_t$,

$$\begin{aligned} P(x_{t+1} \leq x \mid \boldsymbol{z}_t) &= G(x - f(\boldsymbol{z}_t)) \\ &\approx \hat{G}(x - \hat{f}(\boldsymbol{z}_t)) \\ &= \sum_{i=1}^{n} \hat{p}_i I(\tau_i \leq x - \hat{f}(\boldsymbol{z}_t)), \end{aligned} \tag{5.3}$$

where $\hat{G}$ is obtained from the entire fused data $\boldsymbol{\tau}$. From (5.3) we can estimate various conditional functions of $x_{t+1}$ given $\boldsymbol{z}_t$ as byproducts. This procedure is different from methods which use (5.1) directly with only $n_0 << n$ observations.

In practice the $\epsilon_t$ are replaced by the corresponding residuals $\hat{\epsilon}_t$. However, because the residuals of time series regression models are in general dependent, the implementation of the method should proceed with care. One way to tackle the dependence problem is to note that since we are only interested in the distribution of $\hat{\epsilon}_t$, their sequential order is not important. Hence, we can use randomly shuffled or sampled residuals to induce approximate independence, while maintaining the marginal distribution. For example, approximate residual independence may be achieved by using subsequences $\hat{\epsilon}_{t_j}$ where the residuals are spaced sufficiently far apart in time. Nevertheless, using the raw residuals $\hat{\epsilon}_t$ "as is" can lead to useful results as we shall demonstrate in the following examples which illustrate the use of (5.3) in the estimation of the predictive distribution conditional on random covariates. Tacitly, it is assumed that the residuals are equidistributed.

### 5.3.1 Mortality Prediction

Prediction by out of sample fusion is applied here to sampled filtered total mortality data from Los Angeles County, from 01.01.1970 to 12.31.1979 (Shumway et al. 1988). The original daily data, consisting of a response series (total mortality) and its covariate series (two weather and six pollution series), were lowpass filtered (removing frequencies above 0.10 cycles per day) and then sampled weekly to produce series of length $N = 508$ each. Let $y, T, CO$ denote the filtered total mortality, temperature, and carbon monoxide, respectively. A plot of $y_t$ is shown in Figure 5.1, displaying a marked oscillation due to filtering.

It is shown in Kedem and Fokianos (2002) that the regression model

$$y_t = \exp\{\beta_0 + \beta_1 y_{t-1} + \beta_2 y_{t-2} + \beta_3 T_{t-1} + \beta_4 \log(CO_t)\} + \hat{\epsilon}_t \tag{5.4}$$

with partial likelihood estimates

$$(\hat{\beta}_0, \hat{\beta}_1, \hat{\beta}_2, \hat{\beta}_3, \hat{\beta}_4) = (4.5051, 0.0019, 0.0018, -0.0013, 0.0468)$$

and corresponding standard errors (0.0694,0.0004,0.0004,0.0004,0.0087) fits the data well, outperforming several competitors, and giving close to white noise residuals $\hat{\epsilon}_t$ as shown in Figure 5.2. Now, the normal quantile-quantile (qqnorm) plot of the residuals $\hat{\epsilon}_t$ shown in Figure 5.2, as well as the histogram of the residuals shown in Figure 5.3 suggest that fusion of $\hat{\epsilon} \equiv \boldsymbol{x}_0$ with i.i.d. $\eta \equiv \boldsymbol{x}_1 \sim \mathrm{N}(0.0016, 59.5)$ is sensible, where the latter mean and variance are the sample mean and variance of $\hat{\epsilon}_t$. Indeed, with $\boldsymbol{h}(x) = (x, x^2)$ this particular fusion results in $\Delta_{n,obs} = 0.5409109$ which gives a $p$-value of 0.571, legitimizing the density ratio model

$$g_1(x) = \exp\{\alpha + \beta_1 x + \beta_2 x^2\} g(x).$$

Thus, obtaining $\hat{G}$ from the fused data using (4.5) and (4.6) we have,

$$P(y_t > a \mid y_{t-1}, y_{t-2}, T_{t-1}, \log(CO_t)) \approx$$
$$1 - \hat{G}\left(a - \exp\left\{\hat{\beta}_0 + \hat{\beta}_1 y_{t-1} + \hat{\beta}_2 y_{t-2} + \hat{\beta}_3 T_{t-1} + \hat{\beta}_4 \log(CO_t)\right\}\right). \quad (5.5)$$

Confidence intervals for threshold or exceedance probabilities as in (5.5) are discussed in Section 5.4.1.

Figures 5.4 and 5.5 show the predictive probability (5.5) as a function of $t$ for thresholds $a = 180$ and $a = 200$. Apparently, death is more likely to occur during the winter months as manifested clearly in the two figures where the the exceedance probabilities exhibit winter peaks as well as summer troughs. This is supported by the fact that coronary heart disease exhibits a similar seasonal pattern in "incidence and mortality, in countries both north and south of the equator" (Pell and Cobbe 1999).

Identical figures are obtained when $\eta$ is uniform on the residual range, in which case $\Delta_{n,obs} = 0.5188989$ and the $p$-value is 0.382. On the other hand, $\eta \sim \mathrm{Gamma}(3, 2)$ is rejected as $\Delta_{n,obs} = 0.6272471$ with corresponding $p$-value of 0.001.

### 5.3.2 Southern Oscillation Index Prediction

Our next example of out of sample fusion in time series prediction concerns the Southern Oscillation Index (SOI) monthly time series from 1950 to 1995 shown in Figure 5.6. The series consists of 540 monthly observations computed as the difference of the departure from the long-term monthly mean

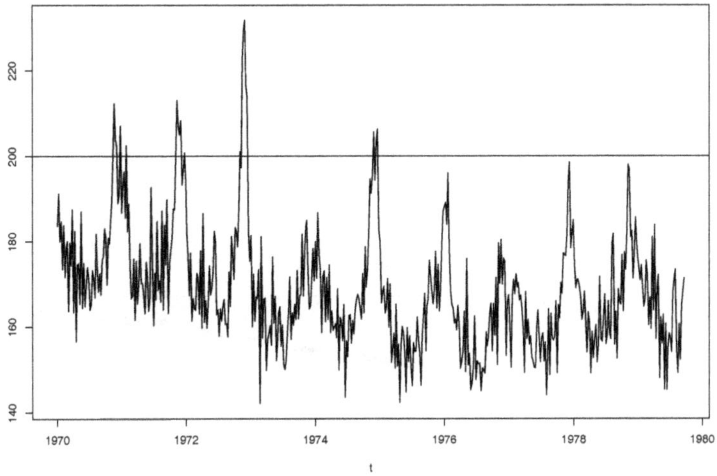


Figure 5.1: Filtered weekly mortality data from Los Angeles County from 01.01.1970 to 12.31.1979 (Shumway et al. 1988).

sea level pressures at Tahiti in the South Pacific and Darwin, Australia. See `https://stat.duke.edu/~mw/data-sets/ts_data/soi`. The SOI is a measure of the large-scale fluctuations in air pressure occurring between the western and eastern tropical Pacific during the so called El Niño and La Niña occurrences. For more on SOI and the methodology used to calculate it see `http://www.ncdc.noaa.gov/teleconnections/enso/indicators/soi/`.

Out of several competitors the AIC criterion points to the following ARMA(2,3) model as optimal,

$$y_t = c + a_1 y_{t-1} + a_2 y_{t-2} + \epsilon_t + b_1 \epsilon_{t-1} + b_2 \epsilon_{t-2} + b_3 \epsilon_{t-3}. \tag{5.6}$$

We observe that since model (5.6) produces residuals, the predictive distribution formula (5.3) is applicable, even though, at least in spirit, ARIMA models differ from the generic model (5.1).

Using the conditional-sum-of-squares to find starting values, then maximum likelihood, the arima() R function gives the estimates

$$(\hat{c}, \hat{a}_1, \hat{a}_2, \hat{b}_1, \hat{b}_2, \hat{b}_3) = (0.0149, 1.7021, -0.7565, -1.2855, 0.3788, 0.1019),$$

with standard errors $(0.1963, 0.0993, 0.0862, 0.1031, 0.0822, 0.0530)$. From Figure (5.7) the residuals $\hat{\epsilon} \equiv \boldsymbol{x}_0$ seem to resemble white noise. Moreover,

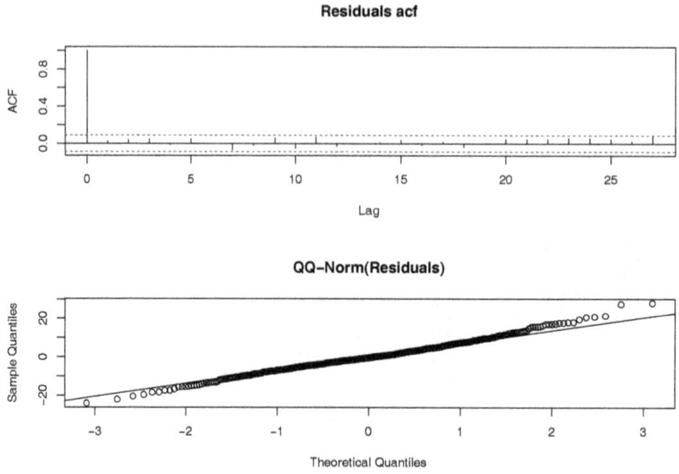


Figure 5.2: Estimated autocorrelation and qqnorm plot of $\hat{\epsilon}_t$ from model (5.4).

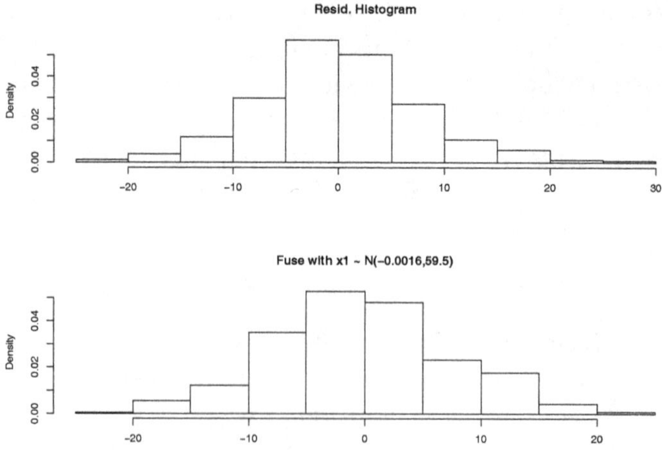


Figure 5.3: Histograms of $\hat{\epsilon} \equiv \boldsymbol{x}_0$ and $\eta \equiv \boldsymbol{x}_1$.

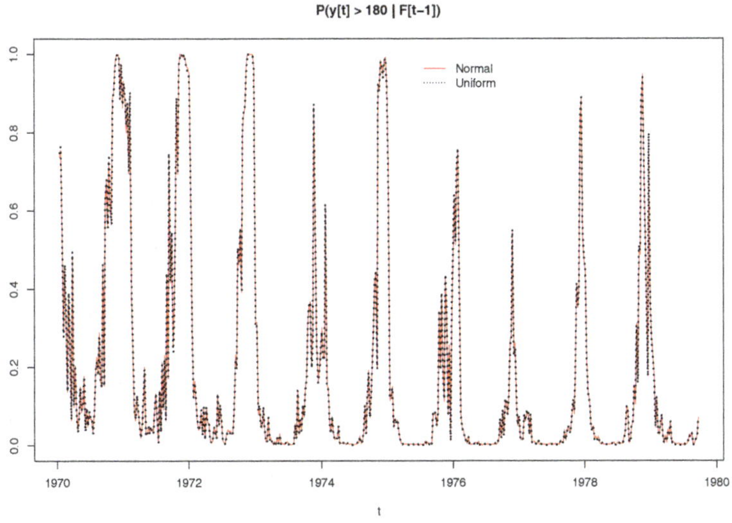


Figure 5.4: $P(y_t > 180 \mid y_{t-1}, y_{t-2}, T_{t-1}, \log(CO_t))$

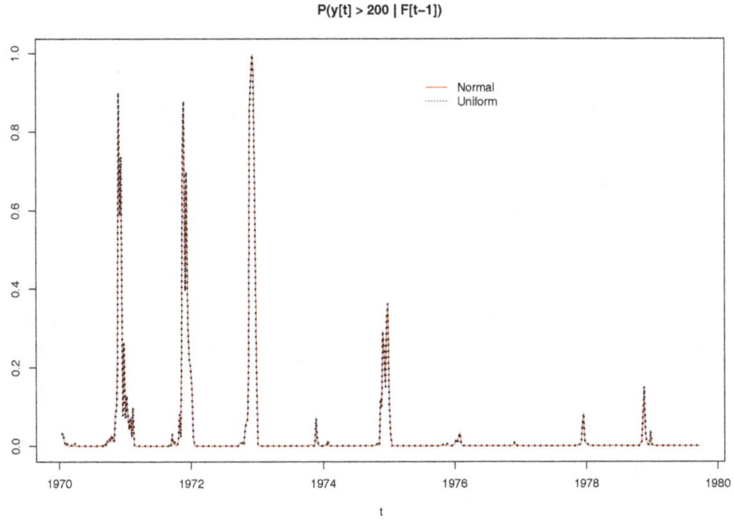


Figure 5.5: $P(y_t > 200 \mid y_{t-1}, y_{t-2}, T_{t-1}, \log(CO_t))$

the normal quantile-quantile plot in Figure (5.7) and the histogram in Figure (5.8) suggest fusing the residuals $\hat{\epsilon} \equiv \boldsymbol{x}_0$ with with i.i.d. $\eta \equiv \boldsymbol{x}_1 \sim \mathrm{N}(\mathrm{mean}(\hat{\epsilon}), \mathrm{var}(\hat{\epsilon}))$, where $(\mathrm{mean}(\hat{\epsilon}), \mathrm{var}(\hat{\epsilon})) = (-0.007524, 1.631841)$, using $h(x) = (x, x^2)$. This in fact gives $\Delta_{n,obs} = 0.5198749$ with corresponding $p$-value of 0.634, lending credence to the density ratio model,

$$g_1(x) = \exp\{\alpha + \beta_1 x + \beta_2 x^2\} g(x).$$

Thus, from (5.3), for any threshold $a$,

$$P(y_t > a \mid y_{t-1}, y_{t-2}, ...) \approx 1 - \hat{G}\left(a - \hat{y}_t\right). \tag{5.7}$$

The future one step exceedance probabilities of thresholds 3 and 6 are shown in Figure 5.9. As the threshold increases, the threshold probability decreases.

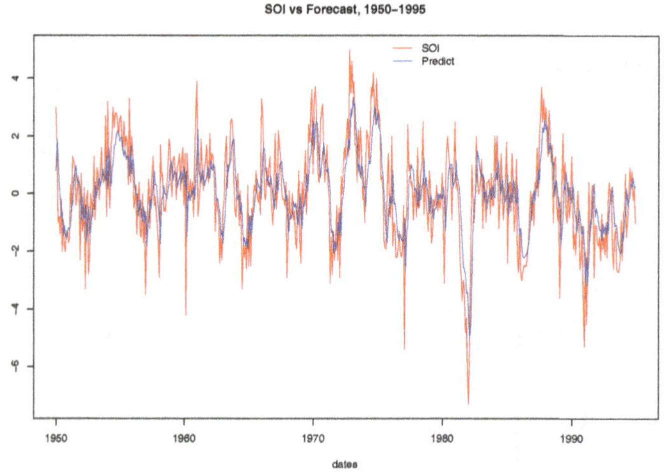


Figure 5.6: Monthly Southern Oscillation Index, 1950-1995, and its one step ahead forecast obtained from the ARMA(2,3) model (5.6).

### 5.3.3 Unemployment Insurance Weekly Claims

The U.S. Employment and Training Administration publishes weekly data on the number of unemployment insurance claims. The data used in the next example consist of seasonally adjusted 4-week moving average of initial claims (IC4WSA). We have used only 400 weekly observations from

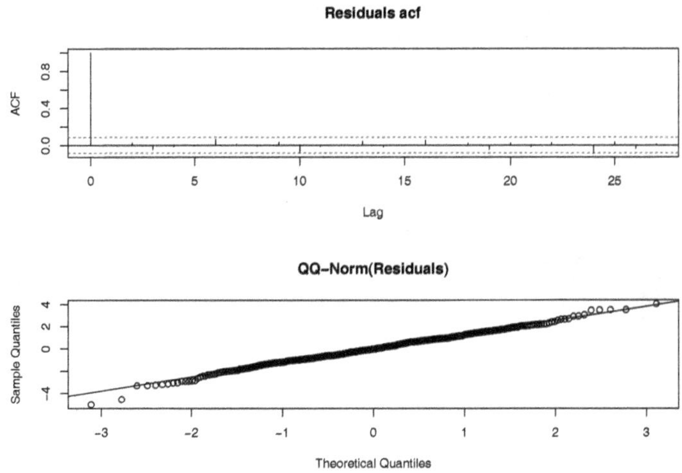


Figure 5.7: Estimated autocorrelation and qqnorm plot of $\hat{\epsilon}_t$ from model (5.6).

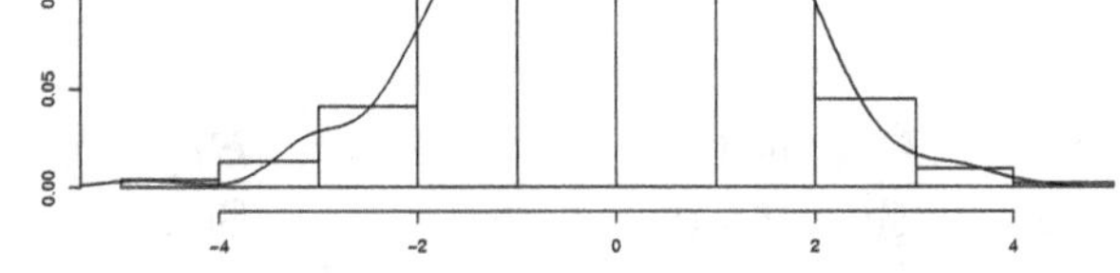

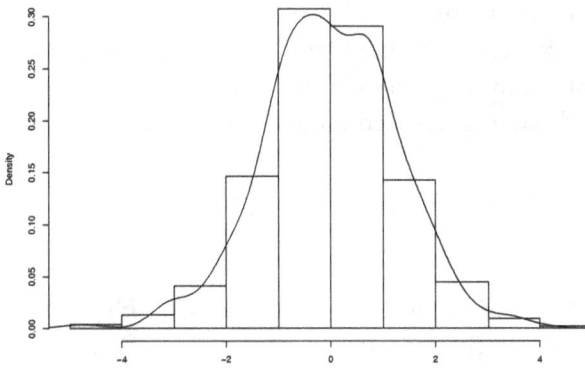

Figure 5.8: Histogram of $\hat{\epsilon} \equiv \boldsymbol{x}_0$ from model (5.6).

2005.01.01 to 2012.08.25 retrieved from the Federal Reserve Bank of St. Louis site `https://research.stlouisfed.org/fred2/series/IC4WSA/`. We wish to predict IC4WSA for the future week of 2012.09.01 which coincides with $t = 401$.

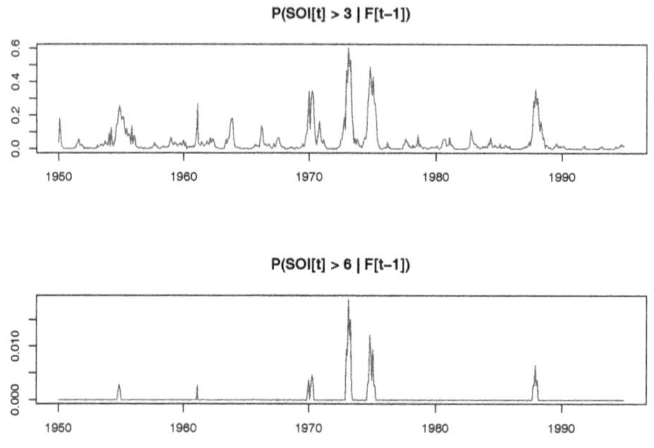


Figure 5.9: $P(y_t > a \mid y_{t-1}, y_{t-2}, ...)$, $a = 3, 6$, for model (5.6).

The first five observations in the data set are 328500, 341750, 344250, 346500, 340250, and the true value for the future date 2012.09.01 is 374250. For computational reasons it is convenient to normalize the observations by $10,000$. The normalized time series is plotted in Figure 5.10 together with its one step ahead forecast obtained from model (5.8) below.

Out of several competing models, the AIC is minimized at the autoregressive integrated moving average model ARIMA(2,1,3),

$$(1 - \sum_{i=1}^{2} a_i \boldsymbol{B}^i) z_t = (1 + \sum_{i=1}^{3} b_i \boldsymbol{B}^i) \epsilon_t \tag{5.8}$$

where $z_t = (1 - \boldsymbol{B}) y_t$, $\boldsymbol{B}$ being the backward shift, $\boldsymbol{B} y_t = y_{t-1}$, and $\epsilon_t$ are independent "shocks" with mean 0 and variance $\sigma_\epsilon^2$ (Box et al 1994, Ch. 4). Applying the same estimation procedure as in the previous SOI example, the parameter estimates are

$$(a_1, a_2, b_1, b_2, b_3) = (-0.2141, -0.2213, 0.9965, 0.9965, 1.0000)$$

with standard errors (0.0495,0.0491,0.0105,0.0223,0.0222).

Following the steps in the previous examples, toward the goal of estimating the reference $G$ needed for the predictive distribution (5.3), our residual analysis is based on 300 residuals, obtained by fitting the model (5.8), sampled at random without replacement out of 400 original residuals. Figure

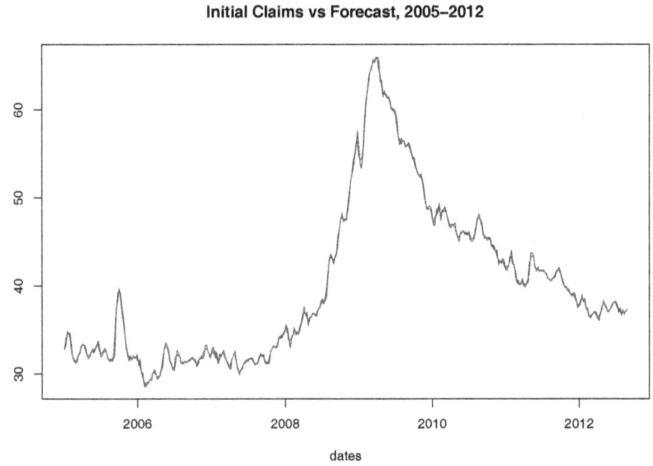


Figure 5.10: 400 observations of 4-week moving average of initial claims (IC4WSA) from 2005.01.01 to 2012.08.25 (red), and its one step ahead forecast (blue) obtained from the ARIMA(2,1,3) model (5.8).

5.11 shows that the sampled residuals are not far from normal white noise, which suggests, again, fusing the residuals $\hat{\epsilon} \equiv \boldsymbol{x}_0$ with i.i.d. $\eta \equiv \boldsymbol{x}_1 \sim \mathrm{N}(\mathrm{mean}(\hat{\epsilon}), \mathrm{var}(\hat{\epsilon}))$, where $(\mathrm{mean}(\hat{\epsilon}), \mathrm{var}(\hat{\epsilon})) = (0.01054406, 0.1402476)$, using $h(x) = (x, x^2)$. This is supported by $\Delta_{n,obs} = 0.4799637$ corresponding to a $p$-value of 0.77. The estimated predictive distribution

$$P(y_t > a \mid y_{t-1}, y_{t-2}, ...) \approx 1 - \hat{G}\left(a - \hat{y}_t\right)$$

is plotted in Figure 5.12. From the estimated predictive distribution we obtain the 95% prediction interval $(36.73, 38.29)$, containing the true future value of 37.425.

### 5.3.4 Forecasting Mortality Rates

The semiparametric prediction method can be ramified in various ways. For example, given a system of time series regression models, we can follow the outline of the previous examples in the estimation of all the predictive distributions by fusing all the residuals from all the regressions, and predict threshold probabilities for each time series using the corresponding $\hat{G}_k$ (Kedem et al. 2008, Kedem and Gagnon 2010). A case in point is the prediction

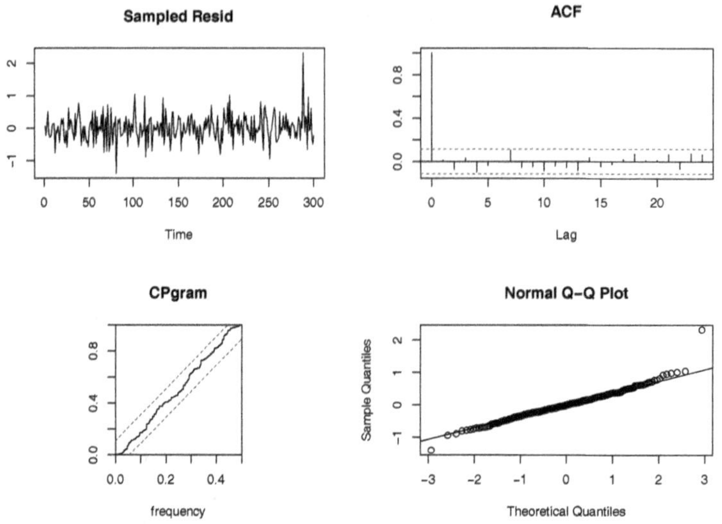


Figure 5.11: Residual analysis corresponding to model (5.8).

of mortality rates in the United States where for each age there is a *short* mortality rate time series represented by a regression model. Unlike the previous examples, here we do not blend the residuals with artificial data although in principle we could do that. In what follows we describe briefly an application to mortality rate prediction discussed in Kedem et al. (2008).

Consider log-death rate $m(k,t)$ for age $k$ and year $t$. Since the U.S. population is large we avoid dealing with zero death-rate at some ages, which is a problem in small states when mortality rates are reported on a log-scale (Voulgaraki et al. 2014). Define $a_k = \sum_t m(k,t)/n$, and let $x_{kt} = m(k,t) - a_k$. For each fixed age $k$, $x_{kt}$ is then the annual time series of centered log death-rate. The system of regression models

$$x_{kt} = b_k x_{k,t-1} + c_k + \epsilon_{kt} \tag{5.9}$$

where $\epsilon_{kt} \sim G_k$ was found appropriate. The drift $c_k$ is added in order to account for a downward trend observed in age-specific mortality rates across all ages in the United States.

The annual mortality rate series for the age group $31-35$ from 1970 to 2001 in the United States, gives five short time series corresponding to ages $31-35$, each of length 32. Therefore, from (5.9), there are now five residual samples for ages $31, 32, 33, 34, 35$, where the residual sample corresponding

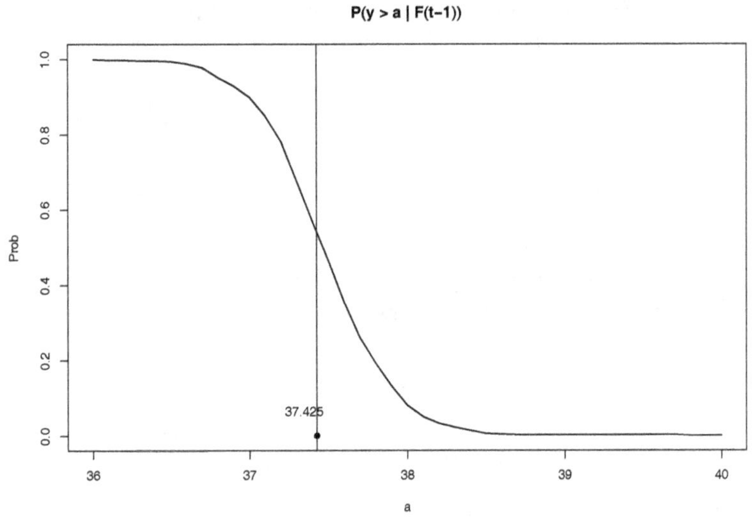


Figure 5.12: $P(y_t > a \mid y_{t-1}, y_{t-2}, ...) \approx 1 - \hat{G}(a - \hat{y}_t)$ corresponding to model (5.8). The true future $y_{t+1} = 37.425$.

to age 33 is taken as the reference sample. Although each residual sample is short, fusing all the residuals gives a total residual sample size of $n = 5 \times 32 = 160$.

We choose a quadratic distortion function $h(x) = x^2$ due to the rough symmetry of the residuals around zero resembling normal residuals with mean zero. Resorting to goodness of fit to support the choice of $h$ in the present application is somewhat problematic as the size of the reference sample is rather small. Combining all the 160 residuals from ages 31 − 35 we get $\hat{G}_k$, $k = 31, ..., 35$, from which, as in (5.3), the predictive distribution for each age is computed by

$$P(x_{k,t+1} \leq x \mid z_{kt}) \approx \hat{G}_k(x - \hat{b}_k x_{kt} - \hat{c}_k) \tag{5.10}$$

where $\hat{b}_k, \hat{c}_k$ are least squares estimates.

We can now apply (5.10) in predicting the probability distribution of the one-year-ahead centered log-death rate in 2002 for each age in the age group 31,32,33,34,35. As a point predictor we use the mean of the predictive distribution, that is, the estimated conditional expectation. The corresponding 95% confidence interval can be derived from the estimated predictive distribution as well. This analysis was repeated for age groups $1-5, 6-10, ..., 81-85$, a total of 17 age groups. A comparison for some ages between the true and predicted (by our method) number of survivors by age and sex out of $100,000$ live births is given in Table 5.1. The true values and their forecasts are close. In this application, the use of (5.10) brings about a significant reduction in the prediction error (Kedem et al. 2008).

Table 5.1: Number of survivors by age and sex, out of 100,000 born alive, from both SP forecasts and true values in 2002.

| | Forecast | | | True | | |
|---|---|---|---|---|---|---|
| Age | Total | Male | Female | Total | Male | Female |
| 0 | 100000 | 100000 | 100000 | 100000 | 100000 | 100000 |
| 1 | 99311 | 99231 | 99371 | 99298 | 99217 | 99360 |
| 10 | 99107 | 99004 | 99186 | 99098 | 98992 | 99184 |
| 20 | 98685 | 98425 | 98914 | 98662 | 98400 | 98902 |
| 30 | 97746 | 97031 | 98394 | 97722 | 97006 | 98384 |
| 40 | 96384 | 95221 | 97425 | 96386 | 95228 | 97422 |
| 50 | 93558 | 91581 | 95293 | 93515 | 91553 | 95220 |
| 60 | 87762 | 84429 | 90632 | 87629 | 84211 | 90570 |
| 70 | 75218 | 69571 | 79978 | 75148 | 69339 | 80074 |
| 80 | 51665 | 43306 | 58630 | 51680 | 43142 | 58758 |

Multi-year ahead forecasts can be obtained by using the predicted values from previous steps. Thus in two-year ahead forecasting we use the previous one-year ahead forecasts, and proceed as above. We note that the prediction error is amplified through each additional step. Another way to obtain multi-year ahead forecasts is to regress the present at time $t$ on observed values up to and including $t-j$, $j = 2, 3....$ Thus in the present case, to get two-year ahead mortality forecasts we use (5.9) with the modification that $x_{kt}$ is regressed on $x_{k,t-2}$.

### 5.3.5 Application to VaR Estimation

We consider briefly a method for the estimation of value-at-risk (VaR) of financial returns using the method outlined in the previous subsection. To apply the method to the calculation of VaR, we need two or more datasets of financial returns, considering one of the datasets as a reference, corresponding to the reference probability density. Again, here as in the previous subsection we do not blend the real data with artificial data but we could do just that if we choose to.

Suppose we have two financial series $P_t$ and $\widetilde{P}_t$. Then the corresponding returns $r_t$ and $\widetilde{r}_t$ are relative changes defined as,

$$\begin{aligned} \ln(P_t) &= \ln(P_{t-1}) + r_t \\ \ln(\widetilde{P}_t) &= \ln(\widetilde{P}_{t-1}) + \widetilde{r}_t. \end{aligned} \tag{5.11}$$

When the temporal changes in the financial series are not large, a Taylor series expansion to one term shows that the returns are really percentage changes. We may view $r_t$ and $\widetilde{r}_t$ as the residuals ($\epsilon$'s) of two regression models. In accordance with the above discussion, if $r_t$ has a reference probability density $g(\cdot)$, and $\widetilde{r}_t$ has density $\widetilde{g}(\cdot)$, then we model the relationship between the densities by

$$\widetilde{g}(x) = \exp(\alpha + \beta x^2) g(x) \tag{5.12}$$

where $g(x)$ is the reference probability density. The choice of $h(x) = x^2$, suggested by the normal distribution with mean zero, is supported by the sound comparison results in Guo (2005).

Given data from $r_t$ and $\widetilde{r}_t$, an application of the semiparametric method yields an estimate for $g(x)$, from which we get the VaR (i.e. a quantile) estimates of $r_t$ corresponding to fixed probabilities. From this, under additional assumptions, we can also estimate the conditional predictive distribution of $P_t$.

A well known method estimates VaR using a normal GARCH(1,1) model (see Kedem and Fokianos (2002) and references therein) under which it is assumed that, conditional on the "past history" $F_{t-1}$,

$$r_t | F_{t-1} \sim \mathrm{N}(0, \sigma_t^2),$$

where $\sigma_t^2$, the conditional variance of $r_t$ given $F_{t-1}$, follows the model

$$\sigma_t^2 = \alpha_0 + \alpha_1 r_{t-1}^2 + \beta_1 \sigma_{t-1}^2. \tag{5.13}$$

This assumption provides a method to determine both the estimated volatility series and the quantile (or VaR) for a given probability. That is, the estimated VaR is the quantile of the standard normal distribution times the estimated standard deviation at time $t$ calculated from (5.13).

An extensive simulation reported in Guo (2005) as well as in Kedem et al. (2005) shows that the GARCH model (5.13) is useful for normal or near normal returns. However, for non-normal returns the semiparametric method competes well with GARCH and is a useful alternative. An example of a non-normal return is the so called TGARCH process where $r_t$ follows a threshold model

$$\begin{aligned} r_t &= \sigma_t \epsilon_t \\ \sigma_t^2 &= \begin{cases} \alpha_0 + \alpha_1 r_{t-1}^2 + \beta_1 \sigma_{t-1}^2 & : \ \epsilon_t > 0 \\ \alpha_0 + \alpha_2 r_{t-1}^2 + \beta_1 \sigma_{t-1}^2 & : \ \epsilon_t \le 0. \end{cases} \end{aligned} \tag{5.14}$$

In empirical work, threshold models such as (5.14) are in tune with "leverage effect". The semiparametric method performs well in terms of bias and mean square error for certain parameter values in (5.11), (5.13), and (5.14).

## 5.4 Interval Estimation of Small Tail Probabilities

By small tail probabilities we mean probabilities of the form $p = 1 - G(T)$ for relatively large thresholds $T$. Such probabilities, also referred to as threshold, survival, or exceedance probabilities, are encountered in many fields where risk is involved. For example in dose-response studies where the problem is to quantify and control the effect on humans or animals caused by relatively high dose levels of some drug. The exceedance probabilities associated with high dose levels are in general very small and could serve as a measure of risk or toxicity. Other important examples include food safety and bio-surveillance, where high levels of contaminants deemed unsafe or even hazardous must be controlled by making sure that the exceedance probabilities of relatively high thresholds are sufficiently small.

The inherent problem in the estimation of small tail probabilities is that in many cases the data do not contain exceedingly large values. Technically, this means that the empirical distribution cannot give useful information about risk or exceedance probabilities for high thresholds when the data do not contain exceedingly large values or "successes" in small and moderately large samples. Any attempt to fit parametric models instead could be a daunting task given highly skewed distributions with very long right tails such as the one represented by the histogram in Figure 5.13, unless we know the exact form of the true distribution up to some parameters. Given this situation, in lieu of real data, and in the spirit of the present chapter, we can approach the problem by fusing the given data with related computer generated "fake" data.

It should be emphasized that we do not deal with the study of extreme values and their distributions associated with the likes of extreme weather conditions, extreme floods, extreme insurance claims, and so on. Our interest is in studying tail probabilities of a distribution and not of its offshoot distributions of maxima and minima whose study requires large samples. As is well known, under certain conditions, the distribution of the maximum of a random sample converges to one of three extreme value distributions regardless of the parent distribution. This fact is useful in the estimation of tail or threshold probabilities from data well below a thershold of interest (Hall and Weissman 1997). Our approach, however, is completely different. A comparison between the two approaches has been made in Pan (2016).

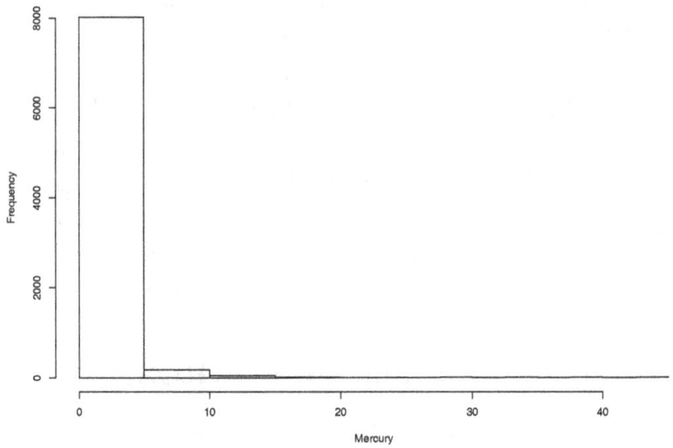


Figure 5.13: Histogram of 8266 methylmercury measurements. Source: http : wwwn.cdc.gov/nchs/nhanes/2007 – 2008/PbCd_E.htm.

### 5.4.1 Confidence Intervals for Tail Probabilities

In moderately large samples, approximate confidence intervals for $p = 1 - G(T)$ for any fixed threshold $T$ can be derived from Theorem 4.1.3.

Denote by $\hat{V}(t)$ the estimated variance of $\hat{G}(t)$ as given in (4.14), obtained by replacing parameters by their estimates. A $1 - \alpha$ level pointwise confidence interval for $G(t)$ is approximated by

$$\left(\hat{G}(t) - z_{\alpha/2}\sqrt{\hat{V}(t)},\ \ \hat{G}(t) + z_{\alpha/2}\sqrt{\hat{V}(t)}\right), \tag{5.15}$$

where $z_{\alpha/2}$ is the upper $\alpha/2$ point of the standard normal distribution. A slight modification of (5.15) gives the desired confidence intervals for tail or survival probabilities $1 - G(T)$ for any $T$, and in particular for relatively large thresholds $T$.

We refer to this interval estimation method of tail probabilities as SP, for semiparametric, and use $\alpha = 0.05$.

The confidence intervals (5.15) are in general shorter than those obtained from nonparametric binomial methods based on the reference sample only. More precisely, by Theorem 4.1.3, due to fusion the precision of the SP confidence intervals is of order $1/\sqrt{n}$, whereas the precision of nonparametric binomial methods based on "success" observations derived from $\boldsymbol{x}_0$ only is $1/\sqrt{n_0}$ where $n_0 << n$.

Interestingly, fusing a given sample with data where the tilt function is misspecified can still, in many cases, lead to useful results as numerous experiments with real and simulated data point out. As an example, suppose we wish to obtain an interval estimate for $p = 1 - G(1.645) = 0.05$ where $\boldsymbol{x}_0 \sim \mathrm{N}(0,1), \boldsymbol{x}_1 \sim \mathrm{N}(1,1), \boldsymbol{x}_2 \sim \mathrm{N}(2,2), \boldsymbol{x}_3 \sim \mathrm{N}(4,4)$. Then for normal samples the tilt function is $h = (x, x^2)$, and with moderate sample sizes we obtain from (5.15) the desired interval with coverage close to the nominal coverage of 95%. Keeping $h = (x, x^2)$, similar coverage is obtained in the misspecified case where the reference $\boldsymbol{x}_0 \sim \mathrm{N}(0,1)$ is fused with non-normal samples $\boldsymbol{x}_1 \sim \mathrm{Exp}(1), \boldsymbol{x}_2 \sim \mathrm{b}(5, 0.6), \boldsymbol{x}_3 \sim \mathrm{Poisson}(1), \boldsymbol{x}_4 \sim \mathrm{t}_{(5)}$.

### 5.4.2 Repeated Out of Sample Fusion (ROSF)

As formulated, the SP method for interval estimation of small tail probabilities requires the density ratio assumption. A surprising fact is that the method often "works" even when the density ratio model assumption does not hold precisely as we saw in the previous example. However, this fact, even if often helpful, does not guarantee acceptable coverage levels in general. Fortunately, to a large degree ROSF fixes this problem as follows.

To motivate the ROSF idea, consider the 95% nominal confidence intervals for $p = 0.01$ in Figure 5.14, where a dark line indicates that the interval does not capture $p$. At the top there are 100 confidence intervals obtained from (5.15) by fusing $\boldsymbol{x}_0 \sim \mathrm{Gamma}(3, 1)$ with $\boldsymbol{x}_1 \sim \mathrm{Unif}(0, 20)$, where $\boldsymbol{h}(x) = (x, \log x)$, $n_0 = n_1 = 100$, and $T = 8.405947$. As the nominal confidence is 95%, and $\boldsymbol{h}(x) = (x, \log x)$ is essentially correct, practically this is a specified case and we would expect 95% coverage. This is indeed the case as seen from the figure. At the bottom, the tilt is still $\boldsymbol{h}(x) = (x, \log x)$ and $n_0 = n_1 = 100$, but $\boldsymbol{x}_0 \sim \mathrm{LN}(1, 1)$, $\boldsymbol{x}_1 \sim \mathrm{Unif}(0, 120)$ , and $T = 27.83649$.

Since the tilt is incorrect, this is a misspecified case, and indeed the coverage is only 72%. However, from the figure, the chance that the upper bounds exceed $p = 0.01$ is *positive*, regardless of the wrong tilt. Likewise, $\boldsymbol{x}_0 \sim \mathrm{f}(2, 12)$ fused with $\boldsymbol{x}_1 \sim \mathrm{Unif}(0, 50)$ gives 83% coverage for $p = 1 - G(T) = 0.01$, and $\boldsymbol{x}_0 \sim \mathrm{Inv-Gaussian}(2, 40)$ fused with $\boldsymbol{x}_1 \sim \mathrm{Unif}(0, 6)$ gives 90%. This fact that the upper bounds exceed $p$ with positive probability using the incorrect Gamma tilt $\boldsymbol{h}(x) = (x, \log x)$ has been observed in many different misspecified cases, and is the basis for ROSF in the estimation of small tail probabilities as described next.

In what follows, since the interest is in very small tail probabilities $p = 1 - G(T)$, the lower bounds of the interval estimates for $p$ are set to zero.

Suppose we wish to estimate a small tail probability $p = 1 - G(T) > 0$ of some distribution, and that a random sample $\boldsymbol{x}_0$ from that distribution is available. We follow a two stage procedure, where first we obtain a certain distribution function, and then use it in the construction of interval estimates for $p$ (Kedem et al. 2016).

Applying OSF for some tilt function $\boldsymbol{h}(x)$, we fuse $\boldsymbol{x}_0$ with a computer generated sample (fusion sample) from some distribution and get from (5.15) a confidence interval $[0, B_1]$ for $p$. We fuse $\boldsymbol{x}_0$ again with another artificial fusion sample, independent of the first and from the same distribution, and get in the same way as before another confidence interval $[0, B_2]$ for $p$. This is repeated many times to produce a sequence of confidence intervals $[0, B_i]$, $i = 1, 2, ..., n$. Since the fusion samples are independent and come from the same distribution, conditional on $\boldsymbol{x}_0$ the upper bounds $B_i$ are now i.i.d. from some distribution $F_B(p)$. Inspired by the previous empirical results, we shall assume that

$$P(B_1 > p) = 1 - F_B(p) > 0. \tag{5.16}$$

The fusions of $\boldsymbol{x}_0$ with computer generated fusion samples can be repeated a large number of times. Therefore, $F_B(p)$ is essentially known for all practical purposes. This follows from the fact that the empirical distribution $\hat{F}_B$ computed from $B_1, B_2, ..., B_n$ converges to $F_B$ almost surely uniformly, as $n$ increases, by the Glivenko-Cantelli Theorem.

**Remark**: Although the distribution of the computer generated fusion samples is known, this fact is not used in our analysis.

Let $B_{(N)}$ be the maximum of a random sample $B_1, ..., B_N$ from $F_B(p)$.

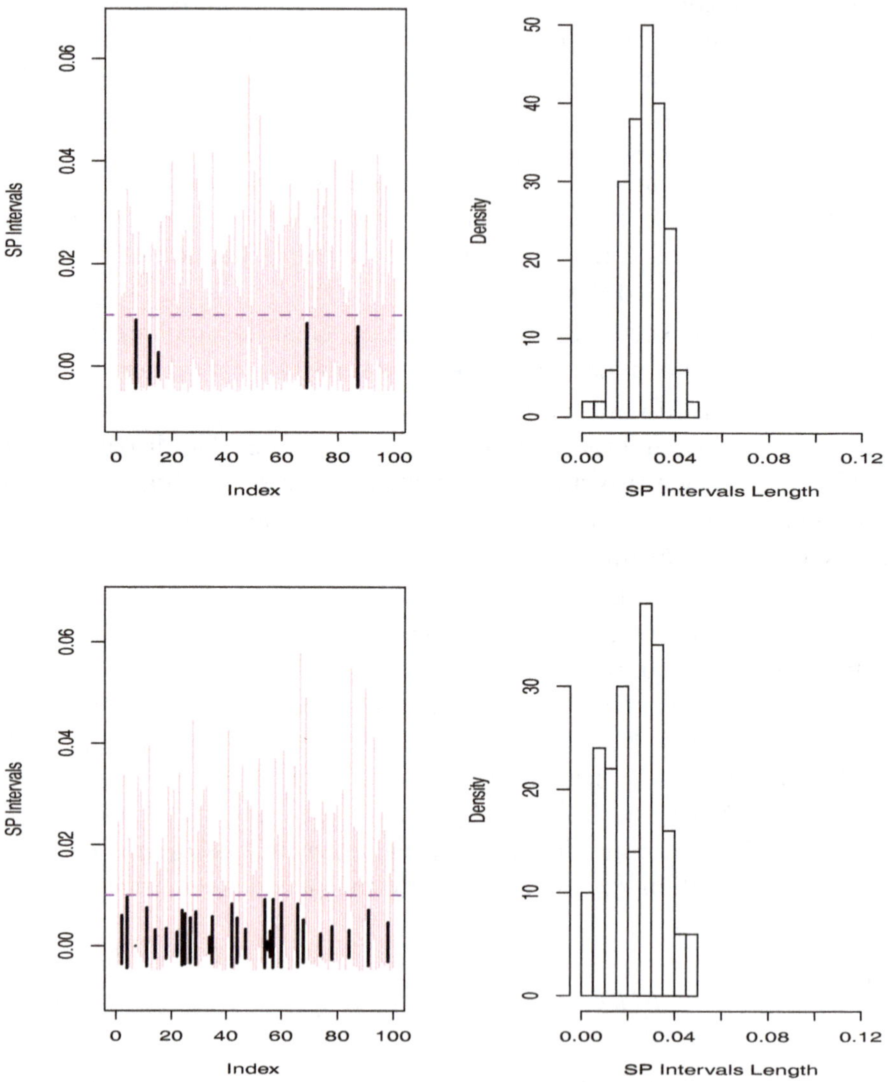


Figure 5.14: 100 nominal 95% OSF confidence intervals for $p = 0.01$. Top: $\boldsymbol{x}_0 \sim \text{Gamma}(3, 1)$, $\boldsymbol{x}_1 \sim \text{Unif}(0, 20)$. Bottom: $\boldsymbol{x}_0 \sim \text{LN}(1, 1)$, $\boldsymbol{x}_1 \sim \text{Unif}(0, 120)$. The histograms represent the 100 widths of the intervals on the left.

Then

$$P(B_{(N)} > p) = 1 - F_B^N(p) \tag{5.17}$$

increases with $N$. It follows that, conditional on the given sample, for all $N > N_0$, for some sufficiently large $N_0$, we have the inequality

$$1 - F_B^N(p) \geq 0.95. \tag{5.18}$$

Hence, the interval

$$0 < p \leq F_B^{-1}(0.05^{1/N}). \tag{5.19}$$

covers $p$ with at least 95% confidence.

From the preceding argument, as long as (5.16) holds and $F_B$ is well defined for some tilt function $\boldsymbol{h}(x)$, ROSF guarantees high coverage of small tail probabilities $p = 1 - G(T)$. Hence, the choice of $\boldsymbol{h}(x)$ is less crucial under ROSF. For skewed data both the gamma $\boldsymbol{h}(x) = (x, \log x)$ and lognormal $\boldsymbol{h}(x) = (\log x, (\log x)^2)$ tilt functions are good practical choices for ROSF.

As mentioned earlier, a detailed comparison between extreme value theory methods, (block maxima and peaks over threshold) and ROSF has been made in Pan (2016). For moderate sample sizes of about 100 the ROSF method gives relatively short intervals as well as high coverage.

### 5.4.3 Illustration of ROSF

Table 5.2 shows simulation results of ROSF for $p = 1 - G(T) = 0.001$, where $\boldsymbol{x}_0$ from different distributions, with very different tail behavior, is fused repeatedly with the indicated computer generated uniform samples $\boldsymbol{x}_1$ using $h(x) = (x, \log x)$. In all cases, $n_0 = n_1 = 100$, and the upper limit of the uniform distribution is chosen to be greater than both $T$ and the largest value of $\boldsymbol{x}_0$.

In each case the table lists only typical upper bounds $F_B^{-1}(0.05^{1/N})$ for various $N$, as the lower bounds for small tail probabilities $p$ are set to 0. In all cases $F_B$ was obtained from $10,000$ $B_i$'s, and the computation time in each case was approximately 12 minutes. The coverage in the table (in bold face) was obtained from 500 runs where in each run $F_B$ was estimated from 200 $B_i$'s only. We see that the coverage increases with $N$ as it should, and that the choice of $N = 100$ seems prudent across all cases, specified or not. However, as we would expect in specified cases, for the two gamma entries at the top of the table, where $h(x) = (x, \log x)$ holds approximately, a much smaller $N = 5$ suffices to achieve high coverage.

We observe that the average $\bar{B}$ from $10,000$ $B_i$'s is not much different from the other entries in the table and thus provides a good idea about the magnitude of $p$.

It is interesting to compare the ROSF results with interval estimation methods of binomial probabilities which only use the clipped data of being above or below a given threshold $T$. In the present simulation, the upper bounds from the method of Agresti and Coull (1998) for $p = 0.001$ and $n_0 = 100$ are almost always 0.0444, 2.5 to 10.5 times larger than the ROSF entries in Table 5.2 corresponding to the conservative $N = 100$, whereas the widespread formula for binomial proportions, known as the Wald interval, gives mostly 0 as an upper bound.

Table 5.2: Typical upper bounds of the confidence intervals (5.19) for $p = 1 - G(T) = 0.001$ for various $N$ across different distributions. $n_0 = n_1 = 100$ and $h(x) = (x, \log x)$. For each typical upper bound, $F_B$ was obtained from $10,000$ $B_i$. In each case, the coverage (in bold face) was obtained from 500 runs of the confidence intervals (5.19). In each run $F_B$ was estimated from 200 $B_i$'s.

| Reference $\boldsymbol{x}_0$ | T | Fusion $\boldsymbol{x}_1$ | $\bar{B}$ | $N = 5$ | $N = 10$ | $N = 20$ | $N = 100$ |
|---|---|---|---|---|---|---|---|
| Gamma(3,1) | 11.229 | Unif(0,15) | 0.0062 | 0.0032 | 0.0039 | 0.0045 | 0.0060 |
| | | | | **0.996** | **0.998** | **1.000** | **1.000** |
| Gamma(1,0.01) | 690.776 | Unif(0,800) | 0.0032 | 0.0032 | 0.0038 | 0.0045 | 0.0059 |
| | | | | **0.990** | **0.994** | **0.996** | **1.000** |
| Pareto(1,4) | 5.623 | Unif(1,8) | 0.0041 | 0.0109 | 0.0119 | 0.0131 | 0.0145 |
| | | | | **0.730** | **0.804** | **0.850** | **0.932** |
| LN(1,1) | 59.754 | Unif(0,140) | 0.0094 | 0.0100 | 0.0126 | 0.0144 | 0.0171 |
| | | | | **0.620** | **0.756** | **0.828** | **0.940** |
| LN(0,1) | 21.982 | Unif(0,60) | 0.0026 | 0.0024 | 0.0038 | 0.0051 | 0.0071 |
| | | | | **0.612** | **0.742** | **0.822** | **0.936** |
| IG(2,40) | 3.836 | Unif(0,6) | 0.0097 | 0.0097 | 0.0116 | 0.0132 | 0.0168 |
| | | | | **0.946** | **0.976** | **0.986** | **0.998** |
| Weibull(0.8,2) | 22.398 | Unif(0,30) | 0.0019 | 0.0019 | 0.0025 | 0.0030 | 0.0042 |
| | | | | **0.928** | **0.964** | **0.974** | **0.992** |

### Application to Lead Exposure

A dataset of 3000 daily intake of lead was constructed by multiplying daily fish consumption by heavy metal concentration (Kedem et al. 2016.) From a histogram of the log-lead exposure data shown in Figure 5.15 we can conclude that the actual data are quite skewed and resemble the mercury data whose histogram is shown in Figure 5.13. In this example, ROSF is applied in the interval estimation of the "true" $p = 1 - G(25) = 0.001$.

Typical upper bounds of the confidence intervals (5.19) for $p = 0.001$ from lead exposure and the corresponding coverage are shown in Table 5.3 for two different fusion distributions. We see that the typical upper bounds in the table for $N \geq 10$ are of the same order as $p$, and that $\bar{B}$ from 10,000 $B_i$'s is somewhat too small, that is, $\bar{B} < p$. Again, as in the previous simulation, we see that the coverage increases with $N$, and that the choice of $N = 100$ is prudent, giving rise to satisfactory coverage. We note, however, attempts to estimate $p = 0.001$ from a sample of size $n_0 = 100$ by fitting a generalized extreme value (GEV) distribution to the lead data proved problematic. Typical GEV estimates were sample-dependent and ranged from 0.01 to 0.03, an order of magnitude greater than $p$.

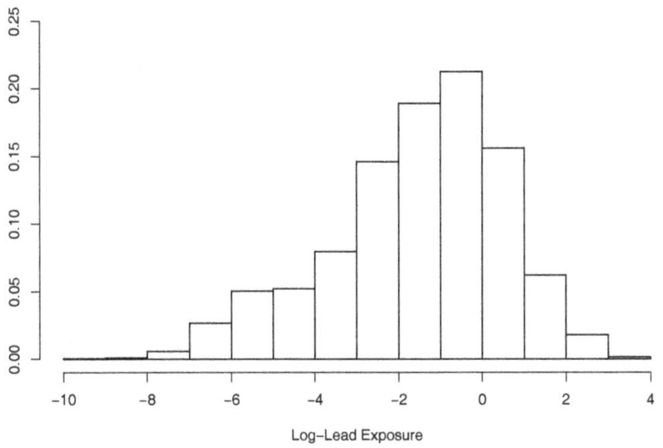


Figure 5.15: Histogram of 3,000 log-lead exposure measurements.

## PSA Application

Total prostate specific antigen (PSA) is a protein produced by the prostate gland and measured in nanograms per milliliter (ng/ml). In healthy men, the PSA level is usually less than 4 ng/ml, and a level exceeding 4 ng/ml is a cause for concern. ROSF is applied next in the estimation of the tail probability that PSA exceeds 50 ng/ml, from a sample of size $n_0 = 100$, drawn randomly from 1,380 total PSA values (Kedem et al. 2016). The data are from the National Health and Nutrition Examination Survey (NHANES)

Table 5.3: Typical upper bounds of the confidence intervals (5.19) for $p = 1 - G(25) = 0.001$ for various $N$ with two different fusion samples using $n_0 = n_1 = 100$ and $h(x) = (x, \log x)$. $F_B$ was obtained from 10,000 $B_i$. The coverage (bold face) was obtained from 500 runs of the confidence intervals (5.19). In each run $F_B$ was estimated from 200 $B_i$'s.

| max($\boldsymbol{x}_0$) | Fusion $\boldsymbol{x}_1$ | $\bar{B}$ | $N=5$ | $N=10$ | $N=20$ | $N=50$ | $N=100$ |
|---|---|---|---|---|---|---|---|
| 15.60 | Unif(1,29) | 0.0007 | 0.0008 | 0.0011 | 0.0013 | 0.0017 | 0.0019 |
| | | | **0.882** | **0.916** | **0.940** | **0.960** | **0.962** |
| 9.59 | Unif(0,60) | 0.0009 | 0.0007 | 0.0013 | 0.0020 | 0.0028 | 0.0033 |
| | | | **0.744** | **0.818** | **0.876** | **0.920** | **0.942** |

collected from men 40 years and older. Among the 1,380 PSA measurements, only the three measurements 52.4,66.6,202.0 exceed 50 ng/ml, and the mode, median, and mean are 0.5, 1.000, and 2.098, respectively, pointing to highly skewed data. See Figure 5.16. Table 5.4 gives typical results corresponding to three different uniform fusion samples. The last upper bound of 0.0013 with $N = 100$ is perhaps too small, apparently a consequence of a relatively small max($\boldsymbol{x}_0$). In this case, a larger $N$ could be entertained.

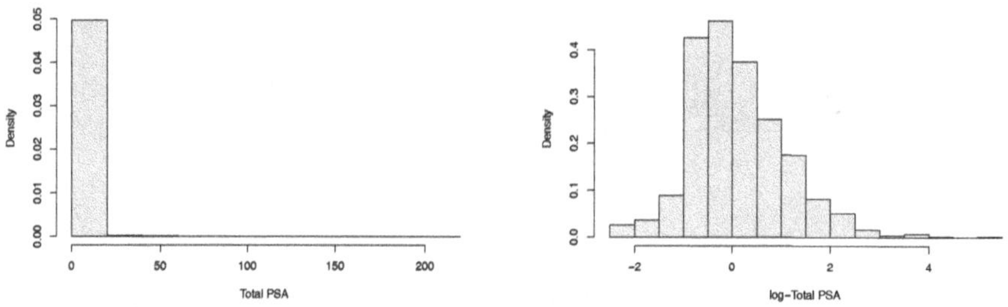


Figure 5.16: Histograms of 1,380 total PSA measurements.

## Application to Methylmercury

The data summarized in Figure 5.13 are that of 8266 measurements of blood methylmercury levels in micrograms per liter obtained from food consumption surveys and measures of methylmercury in fish. Since the data set is large, as well as highly skewed, we can check the applicability of the ROSF method in the estimation of small tail probabilities corresponding to

Table 5.4: Estimation of the probability that total PSA exceeds 50 ng/ml. Typical upper bounds of the confidence intervals (5.19) for $p = 1 - G(50)$ for various $N$ across different fusion samples using $n_0 = n_1 = 100$ and $h(x) = (x, \log x)$. In each entry $F_B$ was obtained from $1,000$ $B_i$.

| $\max(\boldsymbol{x}_0)$ | Fuse | $\bar{B}$ | $N=5$ | $N=10$ | $N=20$ | $N=50$ | $N=100$ |
|---|---|---|---|---|---|---|---|
| 22.2 | Unif(0,55) | 0.0012 | 0.0013 | 0.0018 | 0.0023 | 0.0028 | 0.0034 |
| 14.9 | Unif(0,60) | 0.0006 | 0.0004 | 0.0009 | 0.0014 | 0.0023 | 0.0029 |
| 10.8 | Unif(0,65) | 0.0002 | 0.0001 | 0.0003 | 0.0005 | 0.0010 | 0.0013 |

exceedances of relatively large thresholds, useful in studying the effect of mercury exposure on humans.

The application of ROSF to the interval estimation of, for example, the small tail probability $p = 1 - G(26.5) = 0.0006048875$ with $n_0 = n_1 = 100$, $\boldsymbol{h}(x) = (x, \log x)$, and repeated fusion with Unif(1,80), gives with $N = 100$ 96% coverage and CI mean width of 0.01368543. In this application, taking a clue from the previous simulations, we bypassed knowing $F_B$ and simply took as the interval estimate of $p$ the largest interval $[0, B_{(100)}]$ obtained from 100 fusions. The coverage was obtained from 100 such runs.

# Chapter 6

# Bayesian Inference for Weighted Systems of Distributions

*"The most incomprehensible thing about the world is that it is comprehensible."*
(Albert Einstein, 1879-1955.)

## 6.1 Introduction

The previous chapters illustrate the versatility of the semiparametric density ratio model for combining or fusing statistical information from multiple univariate and multivariate data sources, such as weather measurements from different instruments, microarray data, case-control studies, and time series records. There we used nonparametric and empirical likelihoods within the frequentist inferential approach. In this chapter, however, we describe methodology to deal with some of the problems considered previously by resorting to the Bayesian paradigm, where subjective or context-based prior information is incorporated in the analysis. It seems the semiparametric density ratio model has not been previously analyzed using the Bayesian approach, and in this chapter we propose a possible way of doing that.

There is a vast literature on nonparametric Bayesian inference based on likelihood functions, reviewed for instance in Phadia (2013) and Müller, Quintana, Jara and Hanson (2015), but not much has been done about Bayesian inference based on empirical likelihoods. The reason for this is that, in general, empirical likelihoods may not necessarily be true likelihoods, namely, conditional probability densities when viewed as functions of the data (or a statistic). As a result, the validity of Bayesian inference based on empirical likelihoods is unclear. Monahan and Boos (1992) were the first to investigate the issue of validity of Bayesian inference based on the use of nonnegative integrable functions of the data as likelihoods. They proposed using coverage probability of posterior sets based on the proposed likelihood as a criterion to determine validity, as well as a numerical diagnostic to determine lack of validity. From the latter they found that Bayesian analysis based on some conditional and marginal likelihoods are not valid. Based on this criterion Lazar (2003) provided, for some very simple models, numerical evidence suggesting the (approximate) validity of Bayesian inference based on empirical likelihoods. Although it seems that this issue has not yet been settled formally (but see Schennach, 2005), Lazar's findings have encouraged recent work on Bayesian analysis based on empirical likelihoods. Examples include Chaudhuri and Ghosh (2011), Yang and He (2012), Mengersen, Pudlo and Robert (2013), and Porter, Holan and Wikle (2015).

The Bayesian analysis of the semiparametric density ratio model described here does not rely on an empirical likelihood, as those in the aforementioned works, since estimation of the entire distributions that generate the data is not feasible from the empirical likelihood of the parameters. To answer the questions of interest posed in previous chapters within the context of the semiparametric density ratio model, we need to make inference not only about the 'parametric part' of the model (the $\beta_j$), but also about the 'nonparametric part' ($G$, or more precisely its jumps $p_i$). Because of this, we can work with a certain nonparametric likelihood, similar to those used by Rubin (1981), Chamberlain and Imbens (2003) and Kitamura (2007a,b). This nonparametric likelihood is a true likelihood, and hence the Bayesian analyses that follow from it are always valid. In regard to the prior specification of the 'nonparametric part' of the model, we depart from the practice in previous works which have relied on the Dirichlet distribution, and instead we use a transformed Gaussian distribution. This choice is more flexible as it can represent the type of dependence expected among the $p_i$. The proposed methodology is illustrated by reanalyzing the radar dataset analyzed in Section 2.2.7.

## 6.2 Simple Weighted Systems of Distributions

We shall follow the construction in De Oliveira and Kedem (2017), where an additional quality control application is considered. Let's recall once again the density ratio model studied in Chapter 2 for the case when the function $h(x)$ is real–valued. We have $m = q + 1$ independent random samples following the sampling distributions

$$\begin{aligned}
x_{11}, x_{12}, \ldots, x_{1n_1} &\overset{\text{iid}}{\sim} G_1(x) \\
x_{21}, x_{22}, \ldots, x_{2n_2} &\overset{\text{iid}}{\sim} G_2(x) \\
&\vdots \\
x_{q1}, x_{q2}, \ldots, x_{qn_q} &\overset{\text{iid}}{\sim} G_q(x) \\
x_{m1}, x_{m2}, \ldots, x_{mn_m} &\overset{\text{iid}}{\sim} G(x),
\end{aligned}$$

where the cdfs $G_1, \ldots, G_q$ are exponential distortions of the reference cdf $G$, given by

$$dG_j(x) = \frac{\exp(\beta_j h(x))dG(x)}{\int_{-\infty}^{\infty} \exp(\beta_j h(u))dG(u)}, \qquad j = 1, \ldots, q, \tag{6.1}$$

$\boldsymbol{\beta} = (\beta_1, \ldots, \beta_q)' \in \mathbb{R}^q$ are unknown parameters and $h(x)$ is a known function. The density ratio model studied in Chapter 2 involved parameters $\alpha_j$, but recall that these are functions of $\beta_j$ and $G(x)$ expressed in the denominator of (6.1). For the Bayesian analyses in this chapter it is convenient to work with the above parametrization that only involves the functionally independent parameters $(\boldsymbol{\beta}, G)$. The distributions of all the samples depend on the 'nonparametric' component of the model, $G$, while the distributions of the first $q$ samples also depend on the 'parametric' component, $\boldsymbol{\beta}$ (as well as the known function $h$). As in Chapter 2, we collect all the data from the $m$ independent random samples in a single vector

$$\boldsymbol{t} = (t_1, ..., t_n)' \equiv (\boldsymbol{x}_1', ..., \boldsymbol{x}_q', \boldsymbol{x}_m')',$$

of length $n \equiv n_1 + \cdots + n_q + n_m$, where $\boldsymbol{x}_j = (x_{j1}, ..., x_{jn_j})'$.

We now describe the reference cdf $G$ that would make this semiparametric model flexible. Let $A = \{c_1, c_2, \ldots, c_K\}$ be a finite but large set of points in $\mathbb{R}$, chosen to 'approximate' the support of $G$, and consider the 'nonparametric' family of distributions

$$\mathcal{G} = \Big\{ \sum_{k=1}^{K} p_k I(c_k \le x) : p_k > 0 \text{ for all } k \text{ and } \sum_{k=1}^{K} p_k = 1 \Big\}.$$

It is clear that any distribution defined on the real line can be well approximated by a member of $\mathcal{G}$, provided a reasonable choice of $A$ is used, therefore this family is quite flexible and makes few assumptions. When $G$ belongs to $\mathcal{G}$, then from (6.1) follows that

$$G_j(x) = \sum_{k=1}^{K} \Big( \frac{p_k e^{\beta_j h(c_k)}}{\sum_{l=1}^{K} p_l e^{\beta_j h(c_l)}} \Big) I(c_k \leq x), \qquad j = 1, \ldots, q. \tag{6.2}$$

Then $G \in \mathcal{G}$ and (6.2) provide the specification of the sampling distribution of the semiparametric density ratio model. If $\boldsymbol{p} = (p_1, \ldots, p_n)'$, the likelihood contribution from the reference sample is

$$L(\boldsymbol{\beta}, \boldsymbol{p}; \boldsymbol{x}_m) = \prod_{i=1}^{n_m} \prod_{k=1}^{K} p_k^{I(x_{mi}=c_k)} = \prod_{k=1}^{K} p_k^{N_{mk}},$$

where $N_{mk} = \sum_{i=1}^{n_m} I(x_{mi} = c_k)$ is the number of observations in the reference sample that are equal to the support value $c_k$. Likewise, the likelihood contribution from the $j$th distorted sample is

$$L(\boldsymbol{\beta}, \boldsymbol{p}; \boldsymbol{x}_j) = \prod_{i=1}^{n_j} \prod_{k=1}^{K} \Big( \frac{p_k e^{\beta_j h(c_k)}}{\sum_{l=1}^{K} p_l e^{\beta_j h(c_l)}} \Big)^{I(x_{ji}=c_k)} = \prod_{k=1}^{K} \Big( \frac{p_k e^{\beta_j h(c_k)}}{\sum_{l=1}^{K} p_l e^{\beta_j h(c_l)}} \Big)^{N_{jk}},$$

$j = 1, \ldots, q$ , where $N_{jk} = \sum_{i=1}^{n_j} I(x_{ji} = c_k)$. This is analogous to the sampling model proposed by Rubin (1981), Chamberlain and Imbens (2003) and Ragusa (2014) in their nonparametric Bayesian analyses of one-sample problems. Leonard (1973) proposed a somewhat similar approach involving the nonparametric estimation of a probability density function by a histogram.

### 6.2.1 Likelihood

The works mentioned above provide little or no guidelines about how to choose the set $A$. Here we follow a slight variation of the data-driven choice of $A$ proposed in Kitamura (2007a,b). Let $A = \{t_{(1)}, t_{(2)}, \ldots, t_{(n)}\}$, where the $t_{(k)}$ are the order statistics of the combined sample $\boldsymbol{t}$. It is assumed that $t_{(i)} < t_{(j)}$ for $i < j$ (there are no ties in the data), so $K = n = \sum_{j=1}^{q+1} n_j$. This choice of $A$ is sensible and leads to a flexible model, as long as none of the sample sizes are too small. Now let $\boldsymbol{p} = (\boldsymbol{p}_-', p_n)'$, where $\boldsymbol{p}_- = (p_1, \ldots, p_{n-1})'$, $p_n = 1 - \sum_{k=1}^{n-1} p_k$, and $p_k = dG(t_{(k)})$. Then the semiparametric density ratio model would be parametrized by $(\boldsymbol{\beta}', \boldsymbol{p}_-')' \in$

$\mathbb{R}^q \times \mathbb{S}^{n-1}$, where

$$\mathbb{S}^{n-1} = \{\boldsymbol{p}_- \in \mathbb{R}^{n-1} : p_k > 0 \text{ for all } k \text{ and } \sum_{k=1}^{n-1} p_k < 1\}$$

is the unit simplex in $\mathbb{R}^{n-1}$. For any $k = 1, \ldots, n$ and $j = 1, \ldots, q+1$ it holds that $N_{jk} = 1$ if $t_{(k)}$ belongs to the $j$th sample, and is 0 otherwise. Then the likelihood function of $(\boldsymbol{\beta}', \boldsymbol{p}'_-)'$ based on the $q+1$ samples is

$$\begin{aligned} L(\boldsymbol{\beta}, \boldsymbol{p}_-; \boldsymbol{t}) &= \prod_{k=1}^{n} p_k \cdot \prod_{i=1}^{n_1} \frac{\exp(\beta_1 h(x_{1i}))}{\sum_{l=1}^{n} p_l e^{\beta_1 h(t_{(l)})}} \cdots \prod_{i=1}^{n_q} \frac{\exp(\beta_q h(x_{qi}))}{\sum_{l=1}^{n} p_l e^{\beta_q h(t_{(l)})}} \\ &= \frac{\prod_{k=1}^{n} p_k \cdot \exp(\boldsymbol{\beta}' \boldsymbol{h}_+)}{\left(\sum_{l=1}^{n} p_l e^{\beta_1 h(t_{(l)})}\right)^{n_1} \cdots \left(\sum_{l=1}^{n} p_l e^{\beta_q h(t_{(l)})}\right)^{n_q}} I(\boldsymbol{p}_- \in \mathbb{S}^{n-1}), \end{aligned} \tag{6.3}$$

where $\boldsymbol{\beta} \in \mathbb{R}^q$, $\boldsymbol{p}_- \in \mathbb{S}^{n-1}$, and $\boldsymbol{h}_+ = \left(\sum_{i=1}^{n_1} h(x_{1i}), \ldots, \sum_{i=1}^{n_q} h(x_{qi})\right)'$; this agrees with the empirical likelihood (2.9) in Chapter 2 when the $\alpha_j$ are expressed in terms of the $\beta_j$ and $\boldsymbol{p}_-$.

### 6.2.2 Prior

One of the attractive features of the Bayesian approach is its ability to incorporate into the analysis prior information about the unknown quantities, be it subjective or context-based. For instance, the analyst may believe that the data from the $j$th distorted distribution are stochastically larger than the data from the reference distribution. For the density ratio model this means that $\beta_j > 0$ (provided $h(x)$ is increasing), so a prior may be used for which this event is likely. Also, recall that for the density ratio model described above $p_k = dG(t_{(k)})$, where the $t_{(k)}$ are the observed ordered statistics of the combined or fused sample. Provided the combined sample size is not too small, $t_{(k)}$ and $t_{(k+1)}$ should be close to each other for any $k$, and hence it is reasonable to expect that their corresponding probabilities $p_k$ and $p_{k+1}$ are also close to each other. This is a smoothness property. In what follows we describe prior distributions capable of representing these and other prior beliefs.

Following the Bayesian paradigm, we assume that $\boldsymbol{\beta}$ and $\boldsymbol{p}_-$ are independent a priori random quantities. A flexible prior for $\boldsymbol{\beta}$ is the $\mathrm{N}_q(\boldsymbol{b}_0, B_0)$ distribution, where $\boldsymbol{b}_0$ and $B_0$ are given hyperparameters. The specification of a prior for $\boldsymbol{p}_-$ is more involved, and we discuss here two possibilities. The first, used in similar models by Rubin (1981) and Chamberlain and Imbens (2003), is to assume that $\boldsymbol{p}_- \sim \text{Dir}_{n-1}(a_1, \ldots, a_n)$, where $\text{Dir}_{n-1}(\cdot)$ denotes the Dirichlet distribution on $\mathbb{R}^{n-1}$ and $a_k > 0$ are given hyperparameters. In this case

$$\pi(\boldsymbol{\beta}, \boldsymbol{p}_-) \propto \exp\Big(-\frac{1}{2}(\boldsymbol{\beta} - \boldsymbol{b}_0)' B_0^{-1}(\boldsymbol{\beta} - \boldsymbol{b}_0)\Big) \cdot \prod_{k=1}^{n} p_k^{a_k - 1} I(\boldsymbol{p}_- \in \mathbb{S}^{n-1}), \quad (6.4)$$

and the posterior distribution of $(\boldsymbol{\beta}, \boldsymbol{p}_-)$ based on the $q+1$ samples $\boldsymbol{t}$ is

$$\begin{aligned} \pi(\boldsymbol{\beta}, \boldsymbol{p}_- \mid \boldsymbol{t}) &\propto L(\boldsymbol{\beta}, \boldsymbol{p}_-; \boldsymbol{t}) \pi(\boldsymbol{\beta}, \boldsymbol{p}_-) \\ &= \frac{\exp\big(\boldsymbol{\beta}' \boldsymbol{h}_+ - \frac{1}{2}(\boldsymbol{\beta} - \boldsymbol{b}_0)' B_0^{-1} (\boldsymbol{\beta} - \boldsymbol{b}_0)\big) \cdot \prod_{k=1}^{n} p_k^{a_k}}{\big(\sum_{l=1}^{n} p_l e^{\beta_1 h(t_{(l)})}\big)^{n_1} \cdots \big(\sum_{l=1}^{n} p_l e^{\beta_q h(t_{(l)})}\big)^{n_q}} \\ &\times I(\boldsymbol{p}_- \in \mathbb{S}^{n-1}). \end{aligned} \quad (6.5)$$

A possible default (neutral) prior for the model parameters could be obtained by setting the hyperparameters at $\boldsymbol{b}_0 = \boldsymbol{0}$ and $B_0 = v_0 I_q$, with $v_0 > 0$ and $I_q$ the $q \times q$ identity matrix, and $a_k = a$ for all $k$. The first of these represents the prior belief that all $q+1$ samples come from the same distribution, with $v_0$ controlling the strength of this belief, while the second represents the prior belief that $p_k \sim \text{beta}(a, a(n-1))$, so $E(p_k) = 1/n$ for all $k$. The hyperparameter $a$ quantifies the strength of prior information about the $p_k$, in the sense that increasing $a$ would decrease the prior variance of $p_k$ (O'Hagan, 1994). Unfortunately, the dependence structure of the Dirichlet distribution is quite limited, since it originates from a common denominator present in all the components of the Dirichlet random vector (Aitchison and Shen, 1980; O'Hagan, 1994). In particular, the Dirichlet distribution is inadequate for representing the smoothness property among the $p_k$ described above, since $\text{cov}(p_j, p_k) < 0$ for all $j \neq k$. For these reasons we do not use the Dirichlet prior for $\boldsymbol{p}_-$.

The second possible way to specify a prior for $\boldsymbol{p}_-$ is to assume that a suitable transformation of $\boldsymbol{p}_-$ has a multivariate normal distribution. Several such transformations are possible, some of which were proposed and investigated in Aitchison and Shen (1980) and Aitchison (1986) for several purposes, including Bayesian analysis of contingency tables and frequentist analysis of compositional data. We consider transformations $H : \mathbb{S}^{n-1} \longrightarrow \mathbb{R}^{n-1}$ with the property that each component of $H(\boldsymbol{p}_-)$ is a log-contrast, namely, a variable of the form

$$\sum_{k=1}^{n} c_k \log(p_k), \quad \text{with } \sum_{k=1}^{n} c_k = 0.$$

Each such transformation is one-to-one and has Jacobian proportional to $\prod_{k=1}^{n} p_k^{-1}$ (O'Hagan, 1994). The specific case of such transformations to be used here is

$$H(\boldsymbol{p}_-) = \Big(\log\Big(\frac{p_1}{p_n}\Big), \ldots, \log\Big(\frac{p_{n-1}}{p_n}\Big)\Big)', \tag{6.6}$$

which was studied by Aitchison and Shen (1980). Recall that $p_n = 1 - \sum_{k=1}^{n-1} p_k$, and observe that the $j$th component in (6.6) is the log-contrast obtained by setting $c_j = 1, c_n = -1$ and $c_k = 0$ for $k \neq j, n$. Now, if we assume that $H(\boldsymbol{p}_-) \sim \mathrm{N}_{n-1}(\boldsymbol{m}_0, V_0)$, with given hyperparameters $\boldsymbol{m}_0$ and $V_0$, then the joint pdf of $\boldsymbol{p}_-$ is the so called *logistic-normal* distribution:

$$\pi(\boldsymbol{p}_-) \propto \Big(\prod_{k=1}^{n} p_k\Big)^{-1} \exp\big(-\frac{1}{2}(H(\boldsymbol{p}_-)-\boldsymbol{m}_0)'V_0^{-1}(H(\boldsymbol{p}_-)-\boldsymbol{m}_0)\big)\, I(\boldsymbol{p}_- \in \mathbb{S}^{n-1}). \tag{6.7}$$

For this distribution it holds that for any $j, k = 1, \ldots, n-1$

$$E\Big(\frac{p_j}{p_n}\Big) = \exp\big((\boldsymbol{m}_0)_j + \frac{1}{2}(V_0)_{jj}\big), \tag{6.8}$$

and

$$\mathrm{cov}\Big(\frac{p_j}{p_n}, \frac{p_k}{p_n}\Big) = E\Big(\frac{p_j}{p_n}\Big)E\Big(\frac{p_k}{p_n}\Big)\Big(\exp\big((V_0)_{jk}\big) - 1\Big). \tag{6.9}$$

By still using the marginal prior $\boldsymbol{\beta} \sim \mathrm{N}_q(\boldsymbol{b}_0, B_0)$, the prior $\pi(\boldsymbol{\beta}, \boldsymbol{p}_-)$ is proportional to

$$\begin{aligned} &\Big(\prod_{k=1}^{n} p_k\Big)^{-1} \exp\Big\{-\frac{1}{2}\Big((\boldsymbol{\beta} - \boldsymbol{b}_0)'B_0^{-1}(\boldsymbol{\beta} - \boldsymbol{b}_0) \\ &+ \; (H(\boldsymbol{p}_-) - \boldsymbol{m}_0)'V_0^{-1}(H(\boldsymbol{p}_-) - \boldsymbol{m}_0)\Big)\Big\}\, I(\boldsymbol{p}_- \in \mathbb{S}^{n-1}), \end{aligned} \tag{6.10}$$

and the posterior distribution $\pi(\boldsymbol{\beta}, \boldsymbol{p}_- \mid \boldsymbol{t})$ is proportional to

$$\begin{aligned} &\frac{\exp\Big\{\boldsymbol{\beta}'\boldsymbol{h}_+ - \frac{1}{2}\Big((\boldsymbol{\beta}-\boldsymbol{b}_0)'B_0^{-1}(\boldsymbol{\beta}-\boldsymbol{b}_0) + (H(\boldsymbol{p}_-)-\boldsymbol{m}_0)'V_0^{-1}(H(\boldsymbol{p}_-)-\boldsymbol{m}_0)\Big)\Big\}}{\big(\sum_{l=1}^{n} p_l e^{\beta_1 h(t_{(l)})}\big)^{n_1} \cdots \big(\sum_{l=1}^{n} p_l e^{\beta_q h(t_{(l)})}\big)^{n_q}} \\ &\times I(\boldsymbol{p}_- \in \mathbb{S}^{n-1}). \end{aligned} \tag{6.11}$$

A possible default (neutral) prior for the model parameters in this case is obtained as follows. Again, in (6.10) we set $\boldsymbol{b}_0 = \boldsymbol{0}$ and $B_0 = v_0 I_q$, with $v_0 > 0$. Next, we set $\boldsymbol{m}_0 = -\frac{m_0}{2}\boldsymbol{1}_{n-1}$ and $(V_0)_{jk} = m_0 r_0^{|j-k|}$, with $m_0 > 0$, $\boldsymbol{1}_{n-1}$ a vector of ones, and $r_0 \in (0, 1)$. From (6.8), this implies that $E(p_j/p_n) = 1$ for all $j$, so that $E(p_j) \approx 1/n$ approximately, since $\sum_{j=1}^{n} p_j = 1$, and the approximation improves as $m_0$ decreases to zero. In addition, from (6.9) we have

$$\text{corr}(p_j/p_n, p_k/p_n) = (\exp(m_0 r_0^{|j-k|}) - 1)/(\exp(m_0) - 1) \to 1$$

as $|j-k| \to 0$. Hence, this prior brings about the smoothness property described at the beginning of this section among the ratios $p_k/p_n$, $k = 1, \ldots, n-1$. This in turn implies a similar smoothness property among the $p_k$, since all the ratios share the same denominator. The strength of the smoothness is determined by $r_0$, the closer to 1 $r_0$ is, the more similar $p_k$ and $p_{k+1}$ are expected to be.

We note that a somewhat similar but more involved prior for $\boldsymbol{p}_-$ is proposed in Leonard (1973).

## 6.3 Posterior Simulation

Evidently, the posterior distribution (6.11) is quite non-standard. Consequently, Bayesian inference about $(\boldsymbol{\beta}', \boldsymbol{p}'_-)$ can benefit from the application of Markov chain Monte Carlo (MCMC) methods whose theory and practice are discussed in Robert and Casella (2004) and Gamerman and Lopes (2006). These are Monte Carlo methods that are used when i.i.d. sampling from the posterior distribution of interest is not possible, or practically too cumbersome. The underlying idea is to simulate a Markov chain that has an equilibrium distribution which agrees with the posterior distribution of interest. When this is achieved, any feature of the posterior distribution of interest can be well approximated from the simulated chain, provided that it is long enough and has converged. To make inference about the model parameters we will use a form of Metropolis-Hasting MCMC algorithm in which the parameters are updated separately in two blocks, $\boldsymbol{\beta}$ and $\boldsymbol{p}_-$.

By inspection of (6.11) it follows that the full posterior distributions of $\boldsymbol{\beta}$ is given by

$$\pi(\boldsymbol{\beta} \mid \boldsymbol{p}_-, \boldsymbol{t}) \propto \frac{\exp\left(\boldsymbol{\beta}'\boldsymbol{h}_+ - \frac{1}{2}(\boldsymbol{\beta} - \boldsymbol{b}_0)'B_0^{-1}(\boldsymbol{\beta} - \boldsymbol{b}_0)\right)}{\left(\sum_{l=1}^{n} p_l e^{\beta_1 h(t_{(l)})}\right)^{n_1} \cdots \left(\sum_{l=1}^{n} p_l e^{\beta_q h(t_{(l)})}\right)^{n_q}}.$$

Now, let $(\boldsymbol{\beta}', \boldsymbol{p}'_-)$ denote the current state of the chain, and let $\boldsymbol{\beta}^*$ denote the candidate for the first block of the state in the next iteration. The Metropolis-Hastings update of the first block would be done by first simulating a candidate $\boldsymbol{\beta}^*$ using a random-walk with proposal $q_1(\boldsymbol{\beta}, \boldsymbol{\beta}^*)$ being the $\mathrm{N}_q(\boldsymbol{\beta}, c_1 I_q)$ distribution, where $c_1 > 0$ is a tuning constant. After the candidate $\boldsymbol{\beta}^*$ is simulated, it is accepted with probability given by

$$\alpha_1(\boldsymbol{\beta}, \boldsymbol{\beta}^*) = \min\left\{1, \frac{\pi(\boldsymbol{\beta}^* \mid \boldsymbol{p}_-, \boldsymbol{t}) q_1(\boldsymbol{\beta}^*, \boldsymbol{\beta})}{\pi(\boldsymbol{\beta} \mid \boldsymbol{p}_-, \boldsymbol{t}) q_1(\boldsymbol{\beta}, \boldsymbol{\beta}^*)}\right\} = \min\{1, \xi_1\}, \tag{6.12}$$

where

$$\begin{aligned}\xi_1 &= \left(\frac{\sum_{l=1}^n p_l e^{\beta_1 h(t_{(l)})}}{\sum_{l=1}^n p_l e^{\beta_1^* h(t_{(l)})}}\right)^{n_1} \cdots \left(\frac{\sum_{l=1}^n p_l e^{\beta_q h(t_{(l)})}}{\sum_{l=1}^n p_l e^{\beta_q^* h(t_{(l)})}}\right)^{n_q} \exp\Big\{(\boldsymbol{\beta}^* - \boldsymbol{\beta})'\boldsymbol{h}_+ \\ &- \frac{1}{2}\Big((\boldsymbol{\beta}^* - \boldsymbol{b}_0)' B_0^{-1} (\boldsymbol{\beta}^* - \boldsymbol{b}_0) - (\boldsymbol{\beta} - \boldsymbol{b}_0)' B_0^{-1} (\boldsymbol{\beta} - \boldsymbol{b}_0)\Big)\Big\},\end{aligned}$$

since $q_1(\boldsymbol{\beta}, \boldsymbol{\beta}^*) = q_1(\boldsymbol{\beta}^*, \boldsymbol{\beta})$. If the candidate is not accepted, the next state is set equal to the current state. When the default prior described in Section 6.2.2 is used ($\boldsymbol{b}_0 = \boldsymbol{0}_q$ and $B_0 = v_0 I_q$), the argument in the exponential function above simplifies to

$$\sum_{j=1}^q (\beta_j^* - \beta_j)\Big(\sum_{i=1}^{n_j} h(x_{ji})\Big) - \frac{1}{2v_0}\sum_{j=1}^q (\beta_j^{*2} - \beta_j^2).$$

The guideline for choosing the tuning constant $c_1$ is to set it at a value, chosen by trial and error, that results in an empirical acceptance probability (6.12) in the range 0.25–0.45. This strategy often leads to efficient algorithms when, as in this case, a random-walk Metropolis-Hastings update with a normal proposal is used (Gamerman and Lopes, 2006).

Likewise, it follows from (6.11) that the full posterior distributions of $\boldsymbol{p}_-$ is

$$\pi(\boldsymbol{p}_- \mid \boldsymbol{\beta}, \boldsymbol{t}) \propto \frac{\exp\big(-\frac{1}{2}(H(\boldsymbol{p}_-) - \boldsymbol{m}_0)' V_0^{-1} (H(\boldsymbol{p}_-) - \boldsymbol{m}_0)\big)}{\big(\sum_{l=1}^n p_l e^{\beta_1 h(t_{(l)})}\big)^{n_1} \cdots \big(\sum_{l=1}^n p_l e^{\beta_q h(t_{(l)})}\big)^{n_q}} I(\boldsymbol{p}_- \in \mathbb{S}^{n-1}).$$

Let now $\boldsymbol{p}_-^*$ denote the candidate for the second block of the state in the next iteration. The Metropolis-Hastings update of the second block could be done by simulating a candidate $\boldsymbol{p}_-^*$ using a random-walk with the proposal distribution $q_2(\boldsymbol{p}_-, \boldsymbol{p}_-^*)$ that results by assuming that $H(\boldsymbol{p}_-) \sim \mathrm{N}_{n-1}(\boldsymbol{m}(\boldsymbol{p}_-), c_2 I_{n-1})$ (i.e., $q_2(\boldsymbol{p}_-^*, \cdot)$ is a logistic-normal distribution), with

$$\boldsymbol{m}(\boldsymbol{p}_-) = \log\Big(\frac{1}{p_n}\boldsymbol{p}_-\Big) - \frac{c_2}{2}\boldsymbol{1}_{n-1},$$

where the choice of $\boldsymbol{m}(\boldsymbol{p}_-)$ implies that $E_{q_2}\big(\frac{1}{p_n^*}\boldsymbol{p}_-^*\big) = \frac{1}{p_n}\boldsymbol{p}_-$ (see 6.8). Note that we have used the convention that $\log(a_1, \ldots, a_n) = (\log a_1, \ldots, \log a_n)$.

Unfortunately, this does not result in an efficient algorithm. Exploring this algorithm using several datasets reveals that the simulated draws of $\boldsymbol{p}_-$ are extremely autocorrelated, and so the chain mixes poorly and takes a very large number of iterations to converge. Instead, the candidate for $\boldsymbol{p}_-$ could be simulated using an independence proposal distribution $q_2(\boldsymbol{p}_-, \boldsymbol{p}_-^*)$ being a scaled version of the prior distribution (6.7), meaning that $V_0$ is replaced by $c_2 V_0$, and where $c_2 > 0$ is a tuning constant. After the candidate $\boldsymbol{p}_-^*$ is simulated, it is accepted with probability given by

$$\alpha_2(\boldsymbol{p}_-, \boldsymbol{p}_-^*) = \min\left\{1, \frac{\pi(\boldsymbol{p}_-^* \mid \boldsymbol{\beta}, \boldsymbol{t}) q_2(\boldsymbol{p}_-^*, \boldsymbol{p}_-)}{\pi(\boldsymbol{p}_- \mid \boldsymbol{\beta}, \boldsymbol{t}) q_2(\boldsymbol{p}_-, \boldsymbol{p}_-^*)}\right\} = \min\{1, \xi_2\}, \tag{6.13}$$

where

$$\begin{aligned}\xi_2 &= \prod_{i=1}^n \frac{p_i^*}{p_i} \cdot \left(\frac{\sum_{l=1}^n p_l e^{\beta_1 h(t_{(l)})}}{\sum_{l=1}^n p_l^* e^{\beta_1 h(t_{(l)})}}\right)^{n_1} \cdots \left(\frac{\sum_{l=1}^n p_l e^{\beta_q h(t_{(l)})}}{\sum_{l=1}^n p_l^* e^{\beta_q h(t_{(l)})}}\right)^{n_q} \\ &\quad \times \exp\left\{\left(\frac{c_2 - 1}{2c_2}\right)\Big((H(\boldsymbol{p}_-) - \boldsymbol{m}_0)' V_0^{-1} (H(\boldsymbol{p}_-) - \boldsymbol{m}_0) \\ &\quad -(H(\boldsymbol{p}_-^*) - \boldsymbol{m}_0)' V_0^{-1} (H(\boldsymbol{p}_-^*) - \boldsymbol{m}_0)\Big)\right\}.\end{aligned}$$

If the candidate is not accepted, the next state is set equal to the current state.

When the default prior for $\boldsymbol{p}_-$ described in Section 6.2.2 is used, the evaluation of the above acceptance probability simplifies by the use of two well know identities (that can be checked by inspection). Recall that the matrix $V_0$ involved in the prior of $\boldsymbol{p}_-$ was defined as

$$V_0 = m_0 \begin{pmatrix} 1 & r_0 & r_0^2 & \cdots & r_0^{n-2} \\ r_0 & 1 & r_0 & \cdots & r_0^{n-3} \\ \vdots & & \ddots & & \vdots \\ r_0^{n-2} & r_0^{n-3} & r_0^{n-4} & \cdots & 1 \end{pmatrix}.$$

In this case the inverse of $V_0$ is tridiagonal and given by

$$V_0^{-1} = \frac{1}{m_0(1 - r_0^2)} \begin{pmatrix} 1 & -r_0 & 0 & \cdots & 0 & 0 \\ -r_0 & 1 + r_0^2 & -r_0 & \cdots & 0 & 0 \\ \vdots & & \ddots & & & \vdots \\ 0 & 0 & 0 & \cdots & -r_0 & 1 \end{pmatrix},$$

and for any $\boldsymbol{y} = (y_1, \ldots, y_{n-1})' \in \mathbb{R}^{n-1}$

$$\boldsymbol{y}' V_0^{-1} \boldsymbol{y} = \frac{1}{m_0(1-r_0^2)} \Big( y_1^2 + y_{n-1}^2 + (1+r_0^2) \sum_{i=2}^{n-2} y_i^2 - 2r_0 \sum_{i=1}^{n-2} y_i y_{i+1} \Big). \quad (6.14)$$

Then from (6.6)

$$H(\boldsymbol{p}_-) - \boldsymbol{m}_0 = \Big( \log\Big(\frac{p_1 e^{\frac{m_0}{2}}}{p_n}\Big), \ldots, \log\Big(\frac{p_{n-1} e^{\frac{m_0}{2}}}{p_n}\Big)\Big)',$$

and from (6.14), after some algebraic manipulation, we have

$$\begin{aligned}
\xi_2 \;=\; & \prod_{i=1}^{n} \frac{p_i^*}{p_i} \cdot \left( \frac{\sum_{l=1}^{n} p_l e^{\beta_1 h(t_{(l)})}}{\sum_{l=1}^{n} p_l^* e^{\beta_1 h(t_{(l)})}} \right)^{n_1} \cdots \left( \frac{\sum_{l=1}^{n} p_l e^{\beta_q h(t_{(l)})}}{\sum_{l=1}^{n} p_l^* e^{\beta_q h(t_{(l)})}} \right)^{n_q} \\
& \times \exp\Big\{ \Big( \frac{c_2 - 1}{2c_2 m_0 (1-r_0^2)} \Big) \Big( (1+r_0^2) \sum_{i=2}^{n-2} \Big[ \log^2\Big(\frac{p_i e^{\frac{m_0}{2}}}{p_n}\Big) - \log^2\Big(\frac{p_i^* e^{\frac{m_0}{2}}}{p_n^*}\Big) \Big] \\
& - 2r_0 \sum_{i=1}^{n-2} \Big[ \log\Big(\frac{p_i e^{\frac{m_0}{2}}}{p_n}\Big) \log\Big(\frac{p_{i+1} e^{\frac{m_0}{2}}}{p_n}\Big) - \log\Big(\frac{p_i^* e^{\frac{m_0}{2}}}{p_n^*}\Big) \log\Big(\frac{p_{i+1}^* e^{\frac{m_0}{2}}}{p_n^*}\Big) \Big] \\
& + \log^2\Big(\frac{p_1 e^{\frac{m_0}{2}}}{p_n}\Big) + \log^2\Big(\frac{p_{n-1} e^{\frac{m_0}{2}}}{p_n}\Big) - \log^2\Big(\frac{p_1^* e^{\frac{m_0}{2}}}{p_n^*}\Big) - \log^2\Big(\frac{p_{n-1}^* e^{\frac{m_0}{2}}}{p_n^*}\Big) \Big) \Big\}.
\end{aligned}$$

The guideline for choosing the tuning constant $c_2$ is different from that of $c_1$. Applying this latter algorithm to several datasets reveals that most values of $c_2$ result in chains with extremely low acceptance rates, except when $c_2$ is close to 1. In the latter case, the algorithm produces chains with reasonable acceptance rates that mix well and display relatively small autocorrelations. Empirical results suggest setting $c_2 = 1$ as the best choice, that is, using the prior as a proposal, which not only produces good acceptance rates, but also results in a simplified expression for the acceptance probability (6.13), as part of the argument of the above exponential vanishes. We have found that this results in an efficient algorithm which we will use for the data analysis in the next section. We summarize below the MCMC algorithm to simulate a Markov chain $\{(\boldsymbol{\beta}^{(m)}, \boldsymbol{p}_-^{(m)}) : m = 1, \ldots, M\}$ whose equilibrium distribution is $\pi(\boldsymbol{\beta}, \boldsymbol{p}_- \mid \boldsymbol{t})$.

**Algorithm.**

*Step 1.* Choose the hyperparameters $\boldsymbol{b}_0$, $B_0$, $\boldsymbol{m}_0$, $V_0$, the tuning constants $c_1, c_2$, and the initial state $(\boldsymbol{\beta}^{(0)}, \boldsymbol{p}_-^{(0)})$.

For $m = 1, \ldots, M$ do the following:

*Step 2.* Simulate independently $\boldsymbol{\beta}^* \sim \mathrm{N}_q\big(\boldsymbol{\beta}^{(m-1)}, c_1 I_q\big)$ and $U_1 \sim \mathrm{unif}(0,1)$, and set

$$\boldsymbol{\beta}^{(m)} = \begin{cases} \boldsymbol{\beta}^* & \text{if } U_1 < \alpha_1(\boldsymbol{\beta}^{(m-1)}, \boldsymbol{\beta}^*) \\ \boldsymbol{\beta}^{(m-1)} & \text{otherwise} \end{cases},$$

where $\alpha_1(\cdot,\cdot)$ is given by (6.12).

*Step 3.* Simulate independently $\mathbf{W} = (W_1, \dots, W_{n-1})' \sim \mathrm{N}_{n-1}\big(\boldsymbol{m}_0, c_2 V_0\big)$ and $U_2 \sim \mathrm{unif}(0,1)$, and compute

$$\boldsymbol{p}_-^* = \big(1 + \sum_{i=1}^{n-1} e^{W_i}\big)^{-1}\big(e^{W_1}, \dots, e^{W_{n-1}}\big)'.$$

*Step 4.* Set

$$\boldsymbol{p}_-^{(m)} = \begin{cases} \boldsymbol{p}_-^* & \text{if } U_2 < \alpha_2(\boldsymbol{p}_-^{(m-1)}, \boldsymbol{p}_-^*) \\ \boldsymbol{p}_-^{(m-1)} & \text{otherwise} \end{cases},$$

where $\alpha_2(\cdot,\cdot)$ is given by (6.13), and $p_n^{(m)} = 1 - \mathbf{1}'\boldsymbol{p}_-^{(m)}$.

## 6.4 Bayesian Inference

### 6.4.1 Estimation

Once a large sample $\{(\boldsymbol{\beta}^{(m)}, \boldsymbol{p}_-^{(m)}) : m = 1, \dots, M\}$ from the posterior distribution $\pi(\boldsymbol{\beta}, \boldsymbol{p}_- \mid \boldsymbol{t})$ is available, Bayesian estimates of the quantities of interest follow easily. Point and interval estimates of $\beta_1, \dots, \beta_q$ are constructed from sample averages and quantiles of the corresponding chains. Likewise, a Bayesian estimate of the reference cdf $G$ is given by its posterior expectation

$$\hat{G}^B(x) = E\big(G \mid \boldsymbol{t}\big) = \sum_{k=1}^{n} E(p_k \mid \boldsymbol{t}) I(t_{(k)} \le x),$$

and Bayesian estimates of the distorted cdfs $G_1, \dots, G_q$ are given, using (6.2), by

$$\hat{G}_j^B(x) = E\big(G_j \mid \boldsymbol{t}\big) = \sum_{k=1}^{n} E\Big(\frac{p_k e^{\beta_j h(t_{(k)})}}{\sum_{l=1}^{n} p_l e^{\beta_j h(t_{(l)})}} \mid \boldsymbol{t}\Big) I(t_{(k)} \le x), \quad j = 1, \dots, q,$$

where for each $k$, $E(p_k \mid \boldsymbol{t})$ and $E\Big(\frac{p_k e^{\beta_j h(t_{(k)})}}{\sum_{l=1}^n p_l e^{\beta_j h(t_{(l)})}} \mid \boldsymbol{t}\Big)$ are approximated by sample averages computed from the simulated chain

$$E(p_k \mid \boldsymbol{t}) \approx \frac{1}{M}\sum_{m=1}^{M} p_k^{(m)}$$

$$E\Big(\frac{p_k e^{\beta_j h(t_{(k)})}}{\sum_{l=1}^n p_l e^{\beta_j h(t_{(l)})}} \mid \boldsymbol{t}\Big) \approx \frac{1}{M}\sum_{m=1}^{M} \frac{p_k^{(m)} e^{\beta_j^{(m)} h(t_{(k)})}}{\sum_{l=1}^n p_l^{(m)} e^{\beta_j^{(m)} h(t_{(l)})}}.$$

### 6.4.2 Example: Radar Meteorology

In this section we reanalyze the precipitation radar data described in Section 2.2.7, but now using the Bayesian approach described in this chapter. Recall that the data in question were random samples produced by two different precipitation radars deployed during NASA's Tropical Rainfall Measuring Mission in the Republic of the Marshall Islands. An S-band radar was located on Kwajalein Island, while a C-band radar was located on board of the ship Ronald H. Brown. The two radars produced data to which we apply our Bayesian density ratio method, the goal being the semiparametric estimation of the corresponding Kwajalein and Brown distributions.

The data collection parallels the one used in the analysis in Section 2.2.7. Data from each radar were sampled randomly to produce random samples $\boldsymbol{x}_1$ (Kwajalein, distorted sample) and $\boldsymbol{x}_2$ (Brown, reference sample), each of size 500. Then $q = 1, m = 2, n_1 = n_2 = 500$ and $n = 1000$.

We assume the neutral prior distribution described in Section 6.2.2, that is, $\beta_1$ and $\boldsymbol{p}_-$ are independent a priori, with $\beta_1 \sim N(0, 10)$ ($b_0 = 0$, $v_0 = 10$) and $H(\boldsymbol{p}_-) \sim \mathrm{N}_{999}\big(-0.005, (0.01 \times 0.9^{|j-k|})_{jk}\big)$ ($m_0 = 0.01$, $r_0 = 0.9$), where the transformation $H(\cdot)$ is given in (6.6). First consider the density ratio model with $h(x) = x$. We ran the MCMC algorithm described in Section 6.3, with the tuning constants $c_1 = 0.0003$ and $c_2 = 1$, for $M = 5500$ iterations and a burn-in period of 500. The Metropolis-Hastings updates for $\beta_1$ and $\boldsymbol{p}_-$ had empirical acceptance rates of 0.32 and 0.41, respectively. Figure 6.1 displays a summary of the MCMC output. The first column shows the trace plots of $\beta_1$ and two components of $\boldsymbol{p}_-$, $p_{332}$ and $p_{725}$, and the second column shows the estimated autocorrelation functions of these traces. Together this shows that the algorithm is efficient, as the chain mixes well and has relatively low autocorrelations. The third column shows histogram estimates of the marginal posterior distributions of $\beta_1$, $p_{332}$ and $p_{725}$.

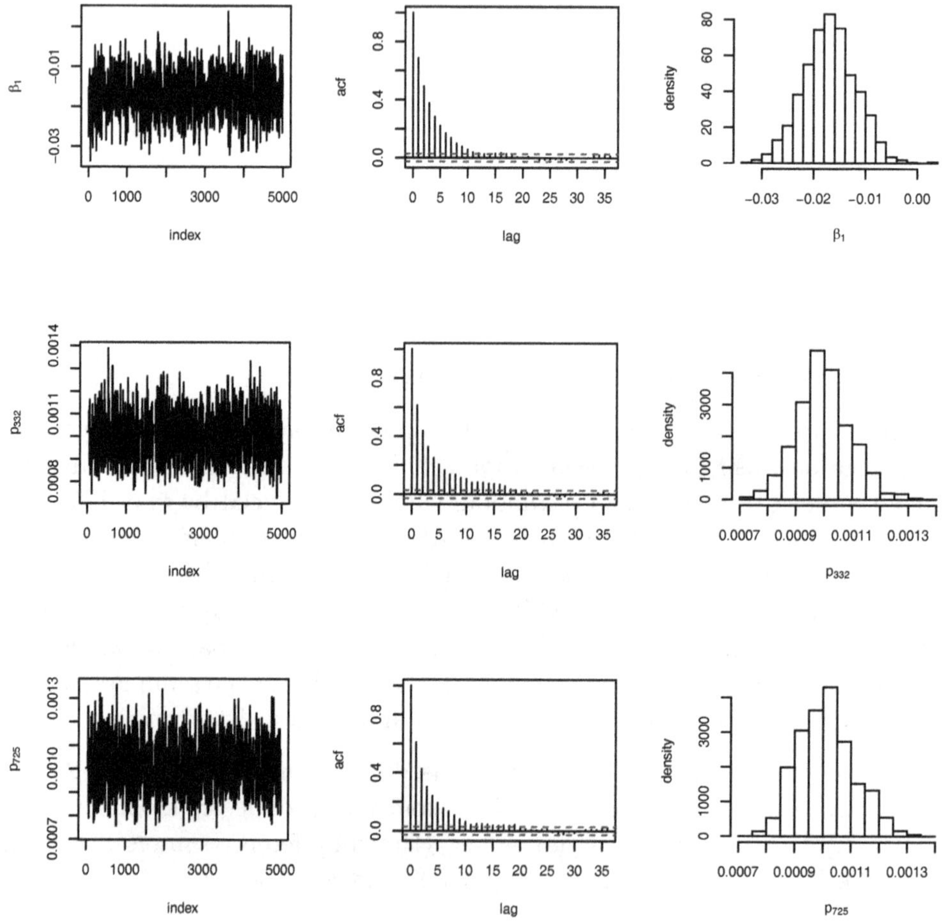


Figure 6.1: Summary of MCMC output of $\beta_1, p_{332}$ and $p_{725}$ from the semi-parametric density ratio model with $h(x) = x$.

The posterior mean of $\beta_1$ is $-0.0171$ and its equal-tail 95% credible interval is $(-0.0268, -0.0073)$. Figure 6.2 displays the estimated cdfs $\hat{G}^B(x)$ and $\hat{G}_1^B(x)$ of Brown's (solid line) and Kwajalein's (dashed line) radar measurements, respectively. It suggests that the two instruments produce precipitation measurements with noticeably different distributions, with Brown's being stochastically larger than Kwajalein's. Note that "B" here stands for "Bayesian" not "Brown".

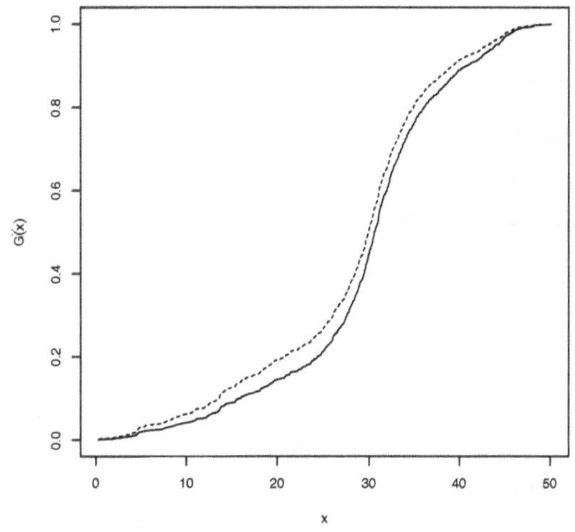


Figure 6.2: Estimates $\hat{G}^B(x)$ (Brown,solid line) and $\hat{G}_1^B(x)$ (Kwajalein, dashed line) from the semiparametric density ratio model with $h(x) = x$.

Next consider the density ratio model with $h(x) = \log(x)$. We again ran the MCMC algorithm for $M = 5500$ iterations and a burn-in period of 500, now with the tuning constants $c_1 = 0.06$ and $c_2 = 1$, which resulted in empirical acceptance rates for $\beta_1$ and $\boldsymbol{p}_-$ of 0.36 and 0.51, respectively. The summary of MCMC output displays an equally good converging and mixing chain as that in Figure 6.1 (not shown). The posterior mean of $\beta_1$ is $-0.2025$ and its equal-tail 95% credible interval is $(-0.3480, -0.0401)$. The substantially different estimate is expected as the interpretation of $\beta_1$ is relative to the chosen function $h(x)$. Figure 6.3 (top) displays the estimated cdfs $\hat{G}^B(x)$ and $\hat{G}_1^B(x)$ of Brown's (solid line) and Kwajalein's (dashed line) radar measurements, respectively. The qualitative conclusion is the same as before, the two instruments produce precipitation measurements with different distributions, but now the two distributions seem to be less distinct. Figure 6.3 (bottom) displays the estimates of the distorted cdfs $\hat{G}_1^B(x)$ (Kwajalein's) under the models with $h(x) = x$ (dashed line) and then with $h(x) = \log(x)$ (solid line), pointing to some sensitivity to the choice of $h(x)$. On the other hand, the estimates of the reference cdf $\hat{G}(x)$ (Brown's), under the model with these two different $h(x)$'s, are visually (and practically) indistinguishable (not shown). Very similar figures were obtained from the frequentist counterpart with a scalar $h(x)$.

### 6.4.3 Test of Hypotheses

As discussed in Chapter 2, one of the main questions of interest when confronted with data from possibly different populations is to assess whether all the samples come from the same population (equidistribution), which in the context of the density ratio model amounts to testing the hypothesis

$$\mathrm{H}_0 : \beta_1 = \cdots = \beta_q = 0,$$

against the alternative $\mathrm{H}_1 : \beta_j \neq 0$ for at least one $j$. In the Bayesian approach, hypothesis testing is usually formulated as a selection problem between two models. Specifically, let $\mathrm{M}_1$ be the Bayesian model specified by the likelihood $L_1(\boldsymbol{\beta}, \boldsymbol{p}_-; \boldsymbol{t})$ in (6.3) and prior $\pi_1(\boldsymbol{\beta}, \boldsymbol{p}_-)$ in (6.10), with $(\boldsymbol{\beta}, \boldsymbol{p}_-) \in \Theta_1 = \mathbb{R}^q \times \mathbb{S}^{n-1}$, and let $\mathrm{M}_0$ be the Bayesian model that results when $\mathrm{H}_0$ holds, so it has likelihood $L_0(\boldsymbol{p}_-; \boldsymbol{t}) = \prod_{k=1}^n p_k \cdot I(\boldsymbol{p}_- \in \mathbb{S}^{n-1})$ and prior $\pi_0(\boldsymbol{p}_-)$ in (6.7), with $\boldsymbol{p}_- \in \Theta_0 = \mathbb{S}^{n-1}$. Testing $\mathrm{H}_0$ versus $\mathrm{H}_1$ is then equivalent to choosing between models $\mathrm{M}_0$ and $\mathrm{M}_1$.

Now, let $\pi_0$ and $\pi_1 = 1 - \pi_0$ be, respectively, the prior probabilities of models $\mathrm{M}_0$ and $\mathrm{M}_1$. The choice between these models would be based on the posterior probability of model $\mathrm{M}_0$, which can be conveniently expressed in terms of the so called *Bayes factor*. The Bayes factor in favor of $\mathrm{M}_0$ is defined as the ratio of posterior to prior odds of $\mathrm{M}_0$, that is

$$\begin{aligned} \mathrm{BF}_{01}(\boldsymbol{t}) &= \frac{P(\mathrm{M}_0 \mid \boldsymbol{t})/(1 - P(\mathrm{M}_0 \mid \boldsymbol{t}))}{\pi_0/(1 - \pi_0)} \\ &= \frac{m_0(\boldsymbol{t})}{m_1(\boldsymbol{t})}, \end{aligned}$$

where the second equality follows from Bayes theorem. The terms $m_0(\boldsymbol{t})$ and $m_1(\boldsymbol{t})$, called the marginal likelihoods (also prior predictive distributions) of

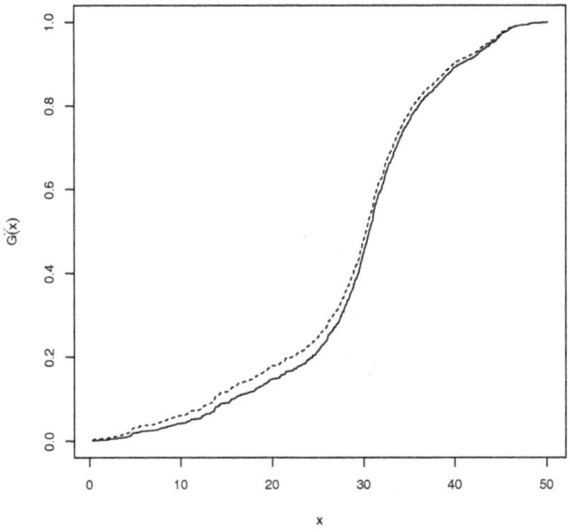


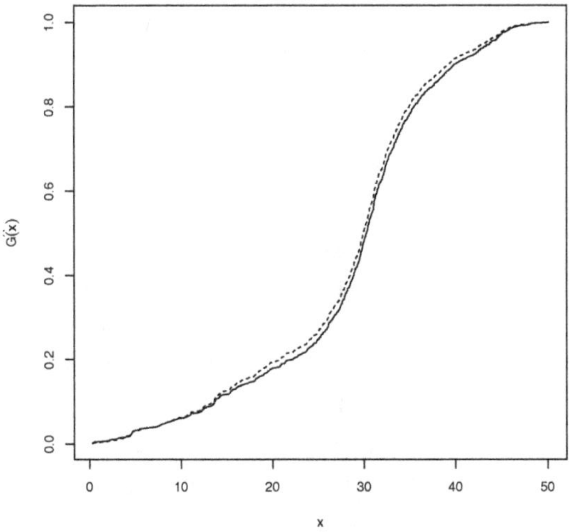


Figure 6.3: Top: Estimates $\hat{G}^B(x)$ (Brown, solid line) and $\hat{G}_1^B(x)$ (Kwajalein, dashed line) from the semiparametric density ratio model with $h(x) = \log(x)$. Bottom: Estimates $\hat{G}_1^B(x)$ from the semiparametric density ratio model with $h(x) = x$ (dashed line) and $h(x) = \log(x)$ (solid line).

the data under models $M_0$ and $M_1$, are given by

$$m_0(\boldsymbol{t}) = \int_{\mathbb{R}^{n-1}} L_0(\boldsymbol{p}_-; \boldsymbol{t})\pi_0(\boldsymbol{p}_-)d\boldsymbol{p}_-$$

and

$$m_1(\boldsymbol{t}) = \int_{\mathbb{R}^q \times \mathbb{R}^{n-1}} L_1(\boldsymbol{\beta}, \boldsymbol{p}_-; \boldsymbol{t})\pi_1(\boldsymbol{\beta}, \boldsymbol{p}_-)d\boldsymbol{\beta}d\boldsymbol{p}_-.$$

The quantities $m_0(\boldsymbol{t})$ and $m_1(\boldsymbol{t})$ are interpreted as the 'evidence' associated with models $M_0$ and $M_1$, respectively, and hence $BF_{01}(\boldsymbol{t})$ is interpreted as the relative evidence in favor of $M_0$ over $M_1$, in light of the observed data, as well as the prior beliefs about the model parameters. A value of $BF_{01}(\boldsymbol{t}) > 1$ points to the conclusion the data lend more support to model $M_0$ than to model $M_1$. This interpretation holds regardless of the prior belief about $M_0$ since $BF_{01}(\boldsymbol{t})$ does not depend on $\pi_0$, which adds to its interpretational appeal. Finally, applying Bayes theorem once more we get

$$P(M_0 \mid \boldsymbol{t}) = \frac{BF_{01}(\boldsymbol{t})\pi_0}{BF_{01}(\boldsymbol{t})\pi_0 + 1 - \pi_0}.$$

The formulation described above shows that the Bayesian approach to testing $H_0$ versus $H_1$ relies on the computation of the marginal likelihoods $m_0(\boldsymbol{t})$ and $m_1(\boldsymbol{t})$, which do not have closed-form expressions for the the density ratio model, and hence they need to be approximated using numerical methods. In general, approximating marginal likelihoods is a notoriously difficult problem, partly due to possibly high dimensional integration, and partly due to the fact that $m_i(\boldsymbol{t})$ is not (naturally) expressible as a posterior functional of $(\boldsymbol{\beta}, \boldsymbol{p}_-)$. Because of this challenge, numerous methods have been proposed in the literature to approximate marginal likelihoods, ranging from numerical quadrature to different Monte Carlo algorithms. Most of these are described in Gamerman and Lopes (2006) and Ando (2010); see also Han and Carlin (2001) for a comparative review. In this respect see Section 3.4.2 dealing with marginalized empirical likelihood applied in tort reform.

For the model comparison considered here the Bayes factor admits an alternative representation that is convenient for computation. Note that model $M_0$ is nested within model $M_1$, since $L_0(\boldsymbol{p}_-; \boldsymbol{t}) = L_1(\boldsymbol{0}, \boldsymbol{p}_-; \boldsymbol{t})$, and in addition $\pi_0(\boldsymbol{p}_-) = \pi_1(\boldsymbol{p}_- \mid \boldsymbol{\beta} = \boldsymbol{0})$, since $\boldsymbol{\beta}$ and $\boldsymbol{p}_-$ are assumed independent a priori under model $M_1$. From these we have, with $f_1$ the conditional pdf of $(\boldsymbol{t}, \boldsymbol{p}_-)$ given $\boldsymbol{\beta} = \boldsymbol{0}$,

$$\begin{aligned} m_0(\boldsymbol{t}) &= \int_{\mathbb{R}^{n-1}} L_0(\boldsymbol{p}_-; \boldsymbol{t}) \pi_0(\boldsymbol{p}_-) d\boldsymbol{p}_- \\ &= \int_{\mathbb{R}^{n-1}} L_1(\boldsymbol{0}, \boldsymbol{p}_-; \boldsymbol{t}) \pi_1(\boldsymbol{p}_- \mid \boldsymbol{\beta} = \boldsymbol{0}) d\boldsymbol{p}_- \\ &= \int_{\mathbb{R}^{n-1}} f_1(\boldsymbol{t}, \boldsymbol{p}_- \mid \boldsymbol{\beta} = \boldsymbol{0}) d\boldsymbol{p}_- \\ &= f_1(\boldsymbol{t} \mid \boldsymbol{\beta} = \boldsymbol{0}) \\ &= \frac{\pi_1(\boldsymbol{\beta} = \boldsymbol{0} \mid \boldsymbol{t}) m_1(\boldsymbol{t})}{\pi_1(\boldsymbol{\beta} = \boldsymbol{0})}, \quad \text{by Bayes theorem,} \end{aligned}$$

where $\pi_1(\cdot)$ and $\pi_1(\cdot \mid \boldsymbol{t})$ are, respectively, the marginal prior and posterior pdfs of $\boldsymbol{\beta}$ under model $\mathrm{M}_1$. Therefore

$$\mathrm{BF}_{01}(\boldsymbol{t}) = \frac{\pi_1(\boldsymbol{\beta} = \boldsymbol{0} \mid \boldsymbol{t})}{\pi_1(\boldsymbol{\beta} = \boldsymbol{0})}, \tag{6.15}$$

the ratio of posterior to prior marginal densities of $\boldsymbol{\beta}$ under model $\mathrm{M}_1$. This latter representation of $\mathrm{BF}_{01}(\boldsymbol{t})$ is called the Savage-Dickey density ratio (O'Hagan, 1994). The denominator of this ratio is available in closed form, while the numerator can be approximated using nonparametric density estimation based on the posterior sample of $\boldsymbol{\beta}$. Note that, by using (6.15), there is no need to fit model $\mathrm{M}_0$ in order to choose between $\mathrm{M}_0$ and $\mathrm{M}_1$.

The practical application and interpretation of Bayes factors for model selection faces, in general, challenges due to their possibly extreme sensitivity to the priors that are chosen for the model parameters. This challenge is specially difficult in problems of selection between nested models, as the one considered here. The nature of the problem becomes apparent from the Bayes factor representation (6.15). For most problems with moderate or large sample sizes the numerator in (6.15) is not very sensitive to the choice of marginal prior $\pi_1(\cdot)$, but the denominator is. Moreover, choosing a 'vague' prior, which works well for estimation problems, would likely result in a large Bayes factor, no matter what the observed data are, so an undue support would be given to model $\mathrm{M}_0$; this is an instance of the so called 'Lindley's paradox' (O'Hagan, 1994). In general this problem requires the use of priors tailored for the goal of model selection, which have not been developed for the present model.

A practical compromise consists of performing model selection using the Bayesian Information Criterion, which provides an approximation for the logarithm of Bayes factors in large samples (under some regularity conditions), and does not depend on the priors for the model parameters (Kass and Raftery, 1995). For the problem of selecting between $M_0$ and $M_1$ this becomes

$$
\begin{aligned}
\log(\mathrm{BF}_{01}(\boldsymbol{t})) &\approx S_{01}(\boldsymbol{t}) \\
&= \log\left(L_0(\hat{\boldsymbol{p}}_-^0;\boldsymbol{t})\right) - \log\left(L_1(\hat{\boldsymbol{\beta}}^1,\hat{\boldsymbol{p}}_-^1;\boldsymbol{t})\right) - \frac{1}{2}(d_0 - d_1)\log(n) \qquad (6.16) \\
&= \left(\frac{q}{2} - n\right)\log(n) - \sum_{k=1}^{n}\log(\hat{p}_k^1) - \boldsymbol{h}_+'\hat{\boldsymbol{\beta}}^1 + \sum_{j=1}^{q} n_j \log\left(\sum_{l=1}^{n}\hat{p}_l^1 e^{\hat{\beta}_j^1 h(t_{(l)})}\right),
\end{aligned}
$$

where $\hat{\boldsymbol{p}}_-^0$ and $(\hat{\boldsymbol{\beta}}^1,\hat{\boldsymbol{p}}_-^1)$ are, respectively, the maximum likelihood estimates under models $M_0$ and $M_1$, and $d_i$ is the number of parameters in model $M_i$, $i = 0, 1$. As shown in Chapter 2, $\hat{\boldsymbol{p}}_-^0 = \frac{1}{n}\mathbf{1}_{n-1}$ and $(\hat{\boldsymbol{\beta}}^1,\hat{\boldsymbol{p}}_-^1)$ is the semiparametric estimator computed in Section 2.3.1.

### 6.4.4 Example: Radar Meteorology (Continuation)

Consider again the precipitation radar data described in Section 6.4.2, and the density ratio model with $h(x) = x$. Using (6.16) we have $S_{01}(\boldsymbol{t}) = -3.0750$, so

$$\mathrm{BF}_{01}(\boldsymbol{t}) \approx e^{-3.0750} = 0.0462 \qquad \text{and} \qquad P(M_0 \mid \boldsymbol{t}) \approx 0.044,$$

where for the latter it is assumed $\pi_0 = 1/2$. Then hypothesis $H_0$ is rejected, and we conclude that the data from Kwajalein and Brown radars come from different distributions. This agrees with the analysis in Section 6.4.2.

The data analysis in Section 6.4.2 showed that the estimate of $G_1(x)$ (the distribution of Kwajalein's radar measurements) was somewhat sensitive to the choice of function $h(x)$. A way of choosing this function is through Bayesian model selection from a set of candidate models with different $h(x)$'s. We illustrate the procedure for the simple case of only two candidate models. Let now $M_1$ be the density ratio model that results from using $h_1(x) = x$, and $M_0$ be the density ratio model that results from using $h_0(x) = \log(x)$. Then, by modifying (6.16) accordingly using (6.3), we have that the last term vanishes, since both models have the same number of parameters, and we obtain $S_{01}(\boldsymbol{t}) = -3.3675$, so in this case

$$\mathrm{BF}_{01}(\boldsymbol{t}) \approx 0.0345 \qquad \text{and} \qquad P(M_0 \mid \boldsymbol{t}) \approx 0.0333,$$

(where again $\pi_0 = 1/2$). We then conclude that these data are better described by the density ratio model with $h_1(x) = x$ than with $h_0(x) = \log(x)$.

## 6.5 Discussion

This chapter describes an approach to Bayesian analysis of the semiparametric density ratio model, in the case when the function $h(x)$ is real–valued. The findings are encouraging, but some remaining issues require further consideration. First, an extension of the proposed methodology using a general tilt function $h(x)$ was not considered. As in the frequentist analysis in Chapter 2, the extension to the case where the function $h(x)$ is vector-valued follows in the footsteps of the scalar case and is straightforward, albeit more involved computationally. Second, it would be desirable to implement the required hypothesis testing in a way that does not rely on large-sample approximations of the Bayes factors. This would require the development of prior distributions tailored to model selection, which presents an interesting research problem. Third, although the proposed MCMC algorithm displayed good performance with the datasets which we explored, it would be desirable to carry out a more formal study of this and other similar algorithms, especially for the analysis of large datasets.

# Chapter 7

# Small Area Estimation

*"Better the illusions that exalt us than ten thousand truths."*
(Alexander Pushkin, 1799-1837.)

This chapter concerns small area estimation under *informative sampling and nonresponse*. As we shall see, there is an interesting connection between the small area problem, as formulated in the present chapter, and the ideas presented hitherto regarding fusing or merging of information, except that here fusion is addressed as "borrowing strength" from areas where information is available. We shall also see examples of weighted distributions and the appeal to computer generated random data presented in previous chapters.

We could approach small area estimation using out of sample fusion as described in Section 5.3, by applying the density ratio model to residuals as done in Kedem et al (2008). In fact, this idea has been exploited recently in Chen and Liu (2016). However, we follow an entirely different path described in recent works of Pfeffermann and Sverchkov listed in the bibliography.

Consider geographical areas such as states, counties, school districts, census regions, socio-economic sub-regions, or any other similar domain. In the context of the present chapter sub-regions are interchangeable with sub-populations. In many cases, a sufficient number of responses is available from a designated area to the extent that reliable inference regarding important area parameters can be carried out using the available area data. Such inference is then referred to as "direct". In many other cases the amount of data from an area is insufficient for direct estimation, in which case we are faced with what is referred to as "small area" estimation. In other words, samples associated with some areas are too small for a meaningful statistical inference. The standard approach to the problem of small area estimation then is to "borrow strength" or use information available from like areas with similar attributes or characteristics, in which case the inference is referred to as "indirect".

More precisely, the problem of small area estimation is to estimate or predict area means, or other quantities of interest, and assess the prediction errors when the area sample sizes are too small (or even zero) to warrant the use of direct design-based estimators. A general approach to small area estimation is based on statistical models which facilitate information borrowing across areas or over time. This procedure is an important example of data fusion.

Comprehensive accounts of small area models and estimation methods can be found in Rao and Molina (2015) and Pfeffermann (2013). The most popular model is the Fay and Herriot (1979) mixed effects area level model. Accordingly, if $\tilde{y}_i$ is a direct sample quantity, such as a sample mean, and $\boldsymbol{x}_i$ is a vector of covariates from area $i$, the model is defined as

$$\tilde{y}_i = \boldsymbol{x}_i'\boldsymbol{\beta} + u_i + e_i$$

where $u_i$ are random effects, and $e_i$ are independent sampling errors with a known covariance matrix. Then $\boldsymbol{\beta}$ is estimated from all the available data from all the areas, usually by generalized least squares, and $\boldsymbol{x}_i'\hat{\boldsymbol{\beta}}$ is referred to as the *synthetic estimator*. The final estimator is a weighted linear combination of the direct and synthetic estimators which minimizes the mean squared error of the estimator under the model. Our aim in this chapter is to consider the so-called *unit level models*, which are in broad use, and moreover are more general than the Fay Herriot model.

If we think of a finite population where the units are spread over a collection of areas, most of the small area models and estimators considered in the literature assume that all the areas are represented in a sample of areas, or that the sampled areas are selected with equal probabilities. A few studies, however, consider the case where the sampling of units within the selected areas is with *unequal selection probabilities* that are related to the outcome values; see Kott (1990), Arora and Lahiri (1997) and Prasad and Rao (1999). These studies only treat the case where the input data consist of the direct estimators of the area means. Kim (2002) considered a unit level model under very special selection model for units within the areas, but it is assumed there implicitly that all areas are selected.

In theory, the effect of the sample selection can be controlled by including among the model covariates all the design variables used for the sample selection. However, this is often not practical either because some or all the design variables may not be known or available at the inference stage, or because there are too many of them, making the fitting and validation of such models formidable. Instead, one could attempt to add to the model the sampling weights as surrogates for the design variables, but the weights may not summarize the information in the design variables adequately, and this proposition is not operational if the sampling weights are not available for the nonsampled areas or units, which is often the case in a secondary analysis. One exception is Verret, Rao and Hidiroglou (2015), to which we refer in Section 7.4.2. As mentioned before, direct design based estimators are highly variable in the sampled areas because of the small sample sizes, and no design based theory exists for the prediction of the means of nonsampled areas, because design based theory uses the randomization distribution of an estimator over repeated sampling from a fixed finite population as the basis for inference. This theory can be used therefore for estimating the population quantities of interest, but not for predicting nonsampled values.

In this chapter we follow the Pfermann and Sverchkov (2007) approach for general unit level models, and illustrate its application in two simple models, the nested area model commonly used for continuous outcomes, and the logistic mixed model suitable for discrete outcomes. We also discuss the underpinning of informative non-response.

The approach deals with situations where the selection of the areas is with unequal probabilities that are possibly related to the true area means, and the sampling of units within the selected areas is with probabilities that are possibly related to the outcome values, even when conditioning on the model covariates. The problem with this kind of sampling designs is that the model holding for the population values no longer holds for the sample data, giving rise to what is known in the sampling literature as informative sampling.

As will be illustrated, failure to account for the effects of informative sampling schemes biases the predictors and increases their root mean squared error. For example, the NHANES III survey, that is used for the empirical application in Section 7.6, oversamples minority groups, and if the target variable of interest (body mass index in our application) is related to ethnicity, then any valid inference procedure should account for the sample selection.

Pfeffermann and Sverchkov (2007) use certain relationships between distributions which hold for the population, the sample, and the unsampled data in order to avoid this problem. To get these relationships one needs to model the selection mechanism, as discussed in Section 7.1.

**Remark concerning design variables:** We shall make repeated reference to design variables. Instead of a formal definition, the following examples suffice. Suppose we wish to estimate the average grade point average (GPA) at some university. Then it is useful to know the number of departments, the number classes taught in each department, and the number of students in each class. These numbers are examples of design variables. In connection with stratified sampling, the indicators of the strata, the number of units in each stratum, and the number of selected units from each stratum, are all design variables. Usually, selection probabilities are functions of design variables.

## 7.1 Sample and Sample-Complement Distributions

Let $A_U = \{(y_{ij}, \boldsymbol{x}_{ij}, \boldsymbol{z}_{ij}); j = 1, ..., N_i; i = 1, ..., M\}$ be a finite population of $N$ units belonging to $M$ areas, with $N_i$ units in area $i$, $y$ is the target variable with values $y_{ij}$ for unit $j$ in area $i$, $\boldsymbol{x}_{ij}$ are the values of corresponding covariates, and $\boldsymbol{z}_{ij}$ are values of design variables. In what follows we consider the population of $y$-values as outcomes of the following two-level random process:

1. *First level* values (*random effects*) $\{u_1, ..., u_M\}$ are generated independently from some distribution with probability density function $f_U(u_i)$. In this chapter we use the abbreviation *pdf* for the probability density function when the random variable is continuous, and probability function when it is discrete.

2. *Second level* values $\{y_{i1}, ..., y_{iN_i}\}$ are generated independently from some distribution with pdf $f_U(y_{ij}|\boldsymbol{x}_{ij}, u_i)$, for $i = 1, ..., M$. By conditioning on random effects we allow for the outcomes to be distributed differently in different areas even after conditioning on known covariates.

We assume a two-stage sampling design by which in the first stage $m$ areas are selected with probabilities $\pi_i = Pr(i \in s|A_U)$, and in the second stage $n_i$ units are sampled from area $i$, selected in the first stage, with probabilities $\pi_{j|i} = Pr(j \in s_i|i \in s, A_U)$, $\pi_i > 0$ and $\pi_{j|i} > 0$ for all $i, j$, where $s$ is the set of selected areas, and $s_i$ is a set of selected units from area $i$. The sample inclusion probabilities are defined by finite population values of design variables $\boldsymbol{z}$ which are known to the sampler who designs the sampling scheme, but not necessary to the analyst. The vector $\boldsymbol{z}$, and therefore the sample inclusion probabilities at both stages, can correlate with the covariates $\boldsymbol{x}$ and the outcome $y$.

**Example:** Let the population model be "unit level random effect model",

$$y_{ij} = \mu + u_i + e_{ij}, \quad u_i \sim N(0, \sigma_u^2), \quad e_{ij} \sim N(0, \sigma_e^2)$$

with the random effects and residual terms being mutually independent. (In this example $\mathbf{x}_{ij} = 1$). Consider the common sampling scheme by which $m$ areas are sampled with probabilities proportional to the area sizes, $\pi_i = mN_i/N$, and the selection within the areas is proportional to some size variable $h_{ij}$ (known to the sampler for the entire population and can correlate with $y_{ij}$), $\pi_{j|i} = n_i h_{ij} / \sum_i^{N_i} h_{ij}$. For example $y_{ij}$ is the current employment in establishment $(i, j)$ and $h_{ij}$ is the last known census employment in that establishment. Then the design variables $\boldsymbol{z}_{ij} = (N_i, h_{ij})'$ define the inclusion probabilities completely.

Denote by $I_i$ and $I_{ij}$ the sample indicator variables for the two sampling stages ($I_i = 1$ iff $i \in s$, and $I_{ij} = 1$ iff $i \in s$ and $j \in s_i$), and by $w_i = 1/\pi_i$ and $w_{j|i} = 1/\pi_{j|i}$ the first and second stage sampling weights.

Following Pfeffermann *et. al* (1998), we define the conditional first level *sample pdf* of $u_i$, that is the *pdf* for ares $i \in s$ as,

$$f_s(u_i) = f_U(u_i | I_i = 1) = Pr(I_i = 1 | u_i) f_U(u_i) / Pr(I_i = 1). \tag{7.1}$$

The conditional first level *sample-complement pdf* of $u_i$ for area $i \notin s$ is defined in Sverchkov and Pfeffermann (2004) as,

$$f_c(u_i) = f_U(u_i | I_i = 0) = Pr(I_i = 0 | u_i) f_U(u_i) / Pr(I_i = 0). \tag{7.2}$$

Note that the *population, sample* and *sample-complement pdfs* of $u_i$ are the same if, $Pr(I_i = 1|u_i) = Pr(I_i = 1)$ for all $i$, in which case the area selection is *noninformative* (or *ignorable*). Otherwise the area selection is *informative* (or *non-ignorable*). Also note that (7.1) and (7.2) are essentially density ratio models.

The conditional second level *sample pdf* and *sample-complement pdfs* of $y_{ij}$ in the sampled area are defined similarly to (7.1) and (7.2) as,

$$\begin{aligned} f_{si}(y_{ij}|\boldsymbol{x}_{ij}, u_i, I_i = 1) &= f_U(y_{ij}|\boldsymbol{x}_{ij}, u_i, I_i = 1, I_{ij} = 1) \\ &= \frac{Pr(I_{ij} = 1|y_{ij}, \boldsymbol{x}_{ij}, u_i, I_i = 1) f_U(y_{ij}|\boldsymbol{x}_{ij}, u_i, I_i = 1)}{Pr(I_{ij} = 1|\boldsymbol{x}_{ij}, u_i, I_i = 1)}, \end{aligned} \tag{7.3}$$

$$\begin{aligned} f_{ci}(y_{ij}|\boldsymbol{x}_{ij}, u_i, I_i = 1) &= f_U(y_{ij}|\boldsymbol{x}_{ij}, u_i, I_i = 1, I_{ij} = 0) \\ &= \frac{Pr(I_{ij} = 0|y_{ij}, \boldsymbol{x}_{ij}, u_i, I_i = 1) f_U(y_{ij}|\boldsymbol{x}_{ij}, u_i, I_i = 1)}{Pr(I_{ij} = 0|\boldsymbol{x}_{ij}, u_i, I_i = 1)}. \end{aligned} \tag{7.4}$$

Here again the *population, sample and sample-complement pdfs* are the same if $Pr(I_{ij} = 1|y_{ij}, \boldsymbol{x}_{ij}, u_i, I_i = 1) = Pr(I_{ij} = 1|\boldsymbol{x}_{ij}, u_i, I_i = 1)$ for all $j$. The model defined by (7.1) and (7.3) is a two-level sample model that corresponds to the population model defined by $f_U(u_i)$ and $f_U(y_{ij}|\boldsymbol{x}_{ij}, u_i)$.

The following relationships between the population pdf, the sample pdf, and the sample-complement pdf are established in Pfeffermann and Sverchkov (1999), and Sverchkov and Pfeffermann (2004), for general pairs of random variables $\mathbf{v}_1$ and $\mathbf{v}_2$ measured for element $i$ (or $ij$) of the population $A_U$. In what follows we use the symbols $E_U$, $E_s$, $E_c$, $E_{si}$ and $E_{ci}$ for the expectations over $f_U$, $f_s$, $f_c$, $f_{si}$ and $f_{ci}$, respectively. Note first that since the inclusion probabilities are defined completely by the finite population $A_U$, then, by the law of iterated expectations,

$$Pr(I_i = 1|\mathbf{v}_{1i}, \mathbf{v}_{2i}) = E_U[E_U(I_i|A_U)|\mathbf{v}_{1i}, \mathbf{v}_{2i}] = E_U(\pi_i|\mathbf{v}_{1i}, \mathbf{v}_{2i}),$$

which implies,

$$\begin{aligned} f_s(\mathbf{v}_{1i}|\mathbf{v}_{2i}) = f_U(\mathbf{v}_{1i}|\mathbf{v}_{2i}, I_i = 1) &= \frac{Pr(I_i = 1|\mathbf{v}_{1i}, \mathbf{v}_{2i}) f_U(\mathbf{v}_{1i}|\mathbf{v}_{2i})}{Pr(I_i = 1|\mathbf{v}_{2i})} \\ &= \frac{E_U(\pi_i|\mathbf{v}_{1i}, \mathbf{v}_{2i}) f_U(\mathbf{v}_{1i}|\mathbf{v}_{2i})}{E_U(\pi_i|\mathbf{v}_{2i})}. \end{aligned} \tag{7.5}$$

Integrating (7.5) gives,

$$E_s(\mathbf{v}_{1i}|\mathbf{v}_{2i}) = \frac{E_U(\pi_i \mathbf{v}_{1i}|\mathbf{v}_{2i})}{E_U(\pi_i|\mathbf{v}_{2i})},$$

and, since $\mathbf{v}_1$ is arbitrary, one can substitute $w_i$, and then $w_i\mathbf{v}_{1i}$ in place of $\mathbf{v}_{1i}$ and get,

$$E_U(\pi_{1i}|\mathbf{v}_{2i}) = \frac{1}{E_s(w_i|\mathbf{v}_{2i})}, \quad E_U(\mathbf{v}_{1i}|\mathbf{v}_{2i}) = \frac{E_s(w_i\mathbf{v}_{1i}|\mathbf{v}_{2i})}{E_s(w_i|\mathbf{v}_{2i})}. \tag{7.6}$$

Note that the left hand side of (7.6) refers to the partially observed data, while the right hand side refers to fully observed data, and therefore the population distribution can be estimated from the observed sample by classical inference such as regression, likelihood or Bayesian methods. Similarly to (7.5) and (7.6) one can get

$$\begin{aligned} f_c(\mathbf{v}_{1i}|\mathbf{v}_{2i}) = f_U(\mathbf{v}_{1i}|\mathbf{v}_{2i}, I_i = 0) &= \frac{E_U[(1-\pi_i)|\mathbf{v}_{1i}, \mathbf{v}_{2i}] f_U(\mathbf{v}_{1i}|\mathbf{v}_{2i})}{E_U[(1-\pi_i)|\mathbf{v}_{2i}]} \\ &= \frac{E_s[(w_i - 1)|\mathbf{v}_{1i}, \mathbf{v}_{2i}] f_s(\mathbf{v}_{1i}|\mathbf{v}_{2i})}{E_s[(w_i - 1)|\mathbf{v}_{2i}]}. \end{aligned} \tag{7.7}$$

$$E_c(\mathbf{v}_{1i}|\mathbf{v}_{2i}) = \frac{E_U[(1-\pi_i)\mathbf{v}_{1i}|\mathbf{v}_{2i}]}{E_U[(1-\pi_i)|\mathbf{v}_{2i}]} = \frac{E_s[(w_i-1)\mathbf{v}_{1i}|\mathbf{v}_{2i}]}{E_s[(w_i-1)|\mathbf{v}_{2i}]}. \tag{7.8}$$

Again, the right hand side of the last equality refers to the observed data, therefore the model for the unobserved data can be estimated from the observed data.

Defining $\mathbf{v}_{1i} = u_i$ and $\mathbf{v}_{2i} = const$ yields the relationships holding for the random area effects $u_i$. Defining $\mathbf{v}_{1ij} = y_{ij}$ and $\mathbf{v}_{2ij} = (\mathbf{x}_{ij}, u_i, I_i = 1)$ and substituting $\pi_{j|i}$ and $w_{j|i}$ for $\pi_i$ and $w_i$, respectively, yields the relationships holding for the observations $y_{ij}$.

## 7.2 Optimal Small Area Predictors

In this chapter we consider estimation of the small area means

$$\bar{Y}_i = \sum_{j=1}^{N_i} y_{ij}/N_i.$$

That is, the means or proportions in sampled and nonsampled areas. Let

$$D_s = \{(y_{ij}, w_{j|i}, w_i), (i,j) \in s; \mathbf{x}_{kl}, (k,l) \in U\}$$

define the known data. Note that we do not assume knowledge of the sampling weights of nonsampled units or areas. The Mean Squared Error (MSE) of a predictor $\hat{\bar{Y}}_i$ with respect to the population pdf, given the observed data $D_s$ and $I_i$ is,

$$\begin{aligned} MSE(\hat{\bar{Y}}_i|D_s, I_i) &= E_U[(\hat{\bar{Y}}_i - \bar{Y}_i)^2|D_s, I_i] \\ &= [\hat{\bar{Y}}_i - E_U(\bar{Y}_i|D_s, I_i)]^2 + V_U(\bar{Y}_i|D_s, I_i). \end{aligned} \tag{7.9}$$

The variance $V_U(\bar{Y}_i|D_s, I_i)$ does not depend on the form of the predictor and hence the optimal predictor that minimizes the MSE in (7.9) is

$$\hat{\bar{Y}}_i = E_U(\bar{Y}_i|D_s, I_i).$$

If area $i$ is sampled ($I_i = 1$), then by (7.7) and the remark below,

$$\begin{aligned} E_U(\bar{Y}_i|D_s, I_i = 1) &= \frac{1}{N_i} E_U\{[\sum_{j=1}^{N_i} E_U(y_{ij}|D_s, u_i, I_i = 1)|D_s]\} \\ &= \frac{1}{N_i}\{\sum_{j\in s_i} y_{ij} + \sum_{l\notin s_i} E_s[E_{ci}(y_{il}|D_s, u_i, I_i = 1)|D_s]\} \\ &= \frac{1}{N_i}\{\sum_{j\in s_i} y_{ij} + \sum_{l\notin s_i} E_s[E_{ci}(y_{il}|\mathbf{x}_{il}, u_i, I_i = 1)|D_s]\}. \end{aligned} \tag{7.10}$$

**Remark.** For the last equality we have to assume that $E_{ci}(y_{il}|D_s, u_i, I_i = 1) = E_{ci}(y_{il}|\mathbf{x}_{il}, u_i, I_i = 1)$. Note that $E_s[E_{ci}(y_{il}|D_s, u_i, I_i = 1)|\mathbf{x}_{il}, u_i, I_i = 1] = E_{ci}(y_{il}|\mathbf{x}_{il}, u_i, I_i = 1)$, so that $E_{ci}(y_{il}|\mathbf{x}_{il}, u_i, I_i = 1)$ 'predicts' $E_{ci}(y_{il}|D_s, u_i, I_i = 1)$ even if this assumption is not satisfied.

For a nonsampled area ($I_i = 0$), if (see below)

$$E_U(y_{ik}|\mathbf{x}_{ik}, u_i, I_i = 0) = E_U(y_{ik}|\mathbf{x}_{ik}, u_i, I_i = 1)$$

then,

$$\begin{aligned}
E_U(\bar{Y}_i|D_s, I_i = 0) &= \frac{1}{N_i} E_U[\sum_{k=1}^{N_i} E_U(y_{ik}|D_s, u_i, I_i = 0)|D_s, I_i = 0] \\
&= \frac{1}{N_i} \sum_{k=1}^{N_i} E_c[E_U(y_{ik}|D_s, u_i, I_i = 0)|D_s] \\
&= \frac{1}{N_i} \sum_{k=1}^{N_i} E_c[E_U(y_{ik}|\mathbf{x}_{ik}, u_i, I_i = 0)|D_s] \\
&= \frac{1}{N_i} \sum_{k=1}^{N_i} E_c[E_U(y_{ik}|\mathbf{x}_{ik}, u_i, I_i = 1)|D_s], \quad (7.11)
\end{aligned}$$

with the first equation on the second line following from the fact that the outcomes $y_{ik}$ are in a nonsampled area. The condition

$$E_U(y_{ik}|\mathbf{x}_{ik}, u_i, I_i = 0) = E_U(y_{ik}|\mathbf{x}_{ik}, u_i, I_i = 1)$$

is not restrictive since the area selection probabilities are related to the area mean and are not dependent on individual deviations from the mean.

## 7.3 Bias of Small Area Predictors when Ignoring an Informative Sampling Scheme

Consider first a sampled area. Ignoring the sampling scheme within a selected area implies an implicit assumption that the sample-complement model in the area is the same as the sample model such that,

$$\bar{Y}_{i,IGN} = 1/N_i\{\sum_{j\in s_i} y_{ij} + \sum_{l\notin s_i} E_s[E_{si}(y_{il}|\mathbf{x}_{il}, u_i, I_i = 1)|D_s]\}$$

(compare with 7.10). Hence

$$
\begin{aligned}
Bias(\bar{Y}_{i,IGN}) &= E_U[(\bar{Y}_{i,IGN} - \bar{Y}_i)|D_s, I_i = 1] \\
&= \frac{1}{N_i}\sum_{l \notin s_i} E_s[E_{si}(y_{il}|\mathbf{x}_{il}, u_i, I_i = 1)|D_s] \\
&- \frac{1}{N_i}\sum_{l \notin s_i} E_s[E_{ci}(y_{il}|\mathbf{x}_{il}, u_i, I_i = 1)|D_s] \\
&= \frac{1}{N_i} E_s[\sum_{l \notin s_i} \frac{Cov_{si}(y_{il}, w_{l|i}|\mathbf{x}_{il}, u_i, I_i = 1)}{E_{si}[(w_{l|i} - 1)|\mathbf{x}_{il}, u_i, I_i = 1)}|D_s]
\end{aligned}
\tag{7.12}
$$

where the last equality follows from (7.8). Thus, if the outcomes $y_{il}$ and the sampling weights $w_{l|i}$ are correlated given the covariates and the random effect, ignoring the sampling scheme yields biased predictors.

Next consider a nonsampled area. By (7.11),

$$
\begin{aligned}
Bias(\bar{Y}_{i,IGN}) &= E_U[(\bar{Y}_{i,IGN} - \bar{Y}_i)|D_s, I_i = 0] \\
&= \frac{1}{N_i}\sum_{k=1}^{N_i} E_s[E_U(y_{ik}|\mathbf{x}_{ik}, u_i, I_{ik} = 1)|D_s] \\
&- \frac{1}{N_i}\sum_{k=1}^{N_i} E_c[E_U(y_{ik}|\mathbf{x}_{ik}, u_i, I_i = 1)|D_s].
\end{aligned}
\tag{7.13}
$$

Adding and subtracting $\frac{1}{N_i}\sum_{k=1}^{N_i} E_c[E_U(y_{ik}|\mathbf{x}_{ik}, u_i, I_{ik} = 1)|D_s]$ and applying (7.8) and (7.6) yields,

$$
\begin{aligned}
Bias(\bar{Y}_{i,IGN}) = &-\frac{1}{N_i}\sum_{k=1}^{N_i} \frac{Cov_s[E_U(y_{ik}|\mathbf{x}_{ik}, u_i, I_{ik} = 1), w_i|D_s]}{E_s[(w_i - 1)|D_s]} \\
&-\frac{1}{N_i} E_c[\sum_{k=1}^{N_i} \frac{Cov_{si}(y_{ik}, w_i|\mathbf{x}_{ik}, u_i, I_i = 1)}{E_{si}(w_{k|i}|\mathbf{x}_{ik}, u_i, I_i = 1)]}|D_s].
\end{aligned}
\tag{7.14}
$$

The first covariance reflects the bias induced by the informative selection of areas. The second covariance reflects the bias induced by the informative sampling within the selected areas (compare with 7.12). In Section 7.5 we propose simple tests for testing whether the covariances in (7.12) and (7.14) are zero, so that ignoring the sample selection does not bias the predictors.

## 7.4 Prediction of Small Area Means

We start with estimation of the model holding for the observed data, $f_s(u_i)$ and $f_{si}(y_{ij}|\boldsymbol{x}_{ij}, u_i, I_i = 1)$. The first step is to fit a parametric model

$$f_{s_\theta}(u_i), \quad f_{si_\theta}(y_{ij}|\boldsymbol{x}_{ij}, u_i, I_i = 1) \tag{7.15}$$

to the observed sample and estimate its (vector) parameter $\theta$ (typicaly generalized mixed linear models are used in small area estimation). Since (7.15) refers to the observed data it can be identified and estimated using standard techniques even when the sampling of areas and within the areas is informative (Rao and Molina 2015).

### 7.4.1 Noninformative Selection of Areas and within the Areas

Now, if the samples of areas and within the areas are noninformative, $f_s(u_i) = f_U(u_i) = f_c(u_i)$ and $f_{si}(y_{ij}|\boldsymbol{x}_{ij}, u_i, I_i = 1) = f_U(y_{ij}|\boldsymbol{x}_{ij}, u_i, I_i = 1) = f_{ci}(y_{ij}|\boldsymbol{x}_{ij}, u_i, I_i = 1)$, then, by (7.10), the *empirical best predictor* $\hat{\bar{Y}}_{i,IGN}$ for the sampled areas, that is, the best predictor $E_U(\bar{Y}_{i,IGN}|D_s, I_i = 1)$, with parameter estimates substituted for model parameters $\boldsymbol{\theta}$, is

$$\begin{aligned}\hat{\bar{Y}}_{i,IGN} &= \hat{E}_U(\bar{Y}_{i,IGN}|D_s, I_i = 1) = \frac{1}{N_i}\{\sum_{j\in s_i} y_{ij} \\ &+ \sum_{l\notin s_i} E_{s_{\hat{\theta}}}[E_{si_{\hat{\theta}}}(y_{il}|\mathbf{x}_{il}, u_i, I_i = 1)|D_s, I_i = 1]\},\end{aligned} \tag{7.16}$$

and by (7.11), the empirical best predictor for nonsampled areas is

$$\begin{aligned}\hat{\bar{Y}}_{i,IGN} &= \hat{E}_U(\bar{Y}_{i,IGN}|D_s, I_i = 0) \\ &= \frac{1}{N_i}\sum_{k=1}^{N_i} E_{s_{\hat{\theta}}}[E_{si_{\hat{\theta}}}(y_{ik}|\mathbf{x}_{ik}, u_i, I_i = 1)|D_s, I_i = 0].\end{aligned} \tag{7.17}$$

**Nested error regression model:** As an example assume that the *nested error regression model*,

$$\begin{aligned} y_{ij} = \mathbf{x}'_{ij}\boldsymbol{\beta} + u_i + e_{ij}; \quad & u_i|I_i = 1 \overset{iid}{\sim} N(0, \sigma_u^2), \\ & e_{ij}|I_{ij} = 1 \overset{iid}{\sim} N(0, \sigma_e^2),\end{aligned} \tag{7.18}$$

fits well the observed data. Under (7.18)

$$u_i|D_s, I_i = 1 \sim N(\hat{u}_i, \sigma_i^2\gamma_i), \tag{7.19}$$

where $\hat{u}_i = \gamma_i[\bar{y}_i - \bar{\mathbf{x}}_i'\boldsymbol{\beta}]$; $(\bar{y}_i, \bar{\mathbf{x}}_i) = \sum_{j=1}^{n_i}(y_{ij}, \mathbf{x}_{ij})/n_i$ are the sample means of $(y, \mathbf{x})$ in sampled area $i$, $\gamma_i = \sigma_u^2/[\sigma_u^2 + \sigma_i^2]$ and $\sigma_i^2 = \sigma_e^2/n_i = Var_s(\bar{y}_i|u_i)$. Then

$$E_s[E_{si}(y_{il}|\mathbf{x}_{il}, u_i, I_i = 1)|D_s, I_i = 1] = \mathbf{x}'_{il}\boldsymbol{\beta} + E_s(u_i|D_s, I_i = 1) = \mathbf{x}'_{il}\boldsymbol{\beta} + \hat{u}_i.$$

Finally, for selected areas,

$$\hat{\bar{Y}}_{i,IGN} = \frac{1}{N_i}[\sum_{j\in s_i} y_{ij} + \sum_{l\notin s_i}(\mathbf{x}'_{il}\hat{\boldsymbol{\beta}} + \hat{u}_i)]. \tag{7.20}$$

Similarly, for nonselected areas,

$$\hat{\bar{Y}}_{i,IGN} = \frac{1}{N_i}\sum_{j=1}^{N_i}\mathbf{x}'_{ij}\hat{\boldsymbol{\beta}}. \tag{7.21}$$

Battese, Harter and Fuller (1988) used the nested error regression model to estimate county crop areas using sample survey data in conjunction with satellite information. In particular, they were interested in estimating the area of corn and soybeans for each of 12 counties in North-Central Iowa. They proposed the model $y_{ij} = \beta_0 + \beta_1 x_{ij1} + \beta_2 x_{ij2} + u_i + e_{ij}$, where $y_{ij} =$ number of hectares of corn (or soybeans), $x_{ij1} =$ number of pixels classified as corn and $x_{ij2} =$ number of pixels classified as soybeans, in the $j$th area segment of farms of the $i$th county. Rao and Choudhry (1995) studied the population of unincorporated tax filers from the province of Nova Scotia, Canada. They proposed the model $y_{ij} = \beta_0 + \beta_1 x_{ij} + u_i + \sqrt{x}_{ij} e_{ij}$, which is similar to (7.19), where $y_{ij}$ and $x_{ij}$ denote the total wages and salaries and gross business income for the $j$th firm in the $i$th area. Simple random sampling from the overall population was used to estimate the small area total $Y_i$ or the means $\bar{Y}_i$.

**Logistic mixed model:** Suppose $y_{ij}$ is binary, that is $y_{ij} = 1$ or 0, and the parameters of interest are the small area proportions, $\bar{Y}_i = \sum_{j=1}^{N_i} y_{ij}/N_i$, and suppose a *logistic mixed model* fits well the observed data,

$$Pr(y_{ij} = 1|\mathbf{x}_{ij}, u_i, I_{ij} = 1) = \frac{\exp(\mathbf{x}'_{ij}\boldsymbol{\beta} + u_i)}{1 + \exp(\mathbf{x}'_{ij}\boldsymbol{\beta} + u_i)}, \quad u_i \overset{iid}{\sim} N(0, \sigma_u^2). \tag{7.22}$$

Then

$$\hat{\bar{Y}}_{i,IGN} = \frac{1}{N_i}[\sum_{j\in s_i} y_{ij} + \sum_{l\notin s_i} \frac{\exp(\mathbf{x'}_{il}\hat{\boldsymbol{\beta}} + \hat{u}_i)}{1+\exp(\mathbf{x'}_{il}\hat{\boldsymbol{\beta}} + \hat{u}_i)}] \tag{7.23}$$

where $\hat{u}_i = \hat{E}_s(u_i|D_s, I_i = 1)$. Note that unlike (7.19), $\hat{u}_i$ does not have a closed form expression.

MacGibbon and Tomberlin (1989) used model (7.22) to estimate area proportions in an empirical study. Malec et al. (1997) applied a model similar to (7.22) to estimate the proportion of persons in a state or substate who have visited a physician in the past year, using data from the U.S. National Health Interview Survey (NHIS).

The most common methods for estimating the regression and variance parameters $\boldsymbol{\beta}$, $\sigma_u^2$, and $\sigma_u^2$ for generalized mixed linear models are maximum likelihood and restricted maximum likelihood. For linear mixed models, the method of moments also can be used. For generalized mixed linear models, Bayesian prediction theory is usually used for predicting the random effect $\hat{u}_i$. The most popular statistical software packages have the capacity to compute all these estimates.

For a review of the use of mixed models for small area estimation, see Datta (2009).

There exists a well developed theory about MSE estimation of small area predictors under noninformative sampling (e.g. Rao and Molina 2015). Unfortunately it does not work directly when the sampling scheme is informative, except for the case considered in Verret, Rao and Hidiroglou (2015) discussed in the following section.

### 7.4.2 Informative Selection withing the Areas when all Areas are Selected

A possible way to account for the sampling effects is to include among the model covariates all the variables and interactions determining the sample probabilities. It requires knowledge of population values of all the variables determining the sample selection. Although the population values of the design variables used for the sample selection are known to the sampler drawing the sample, they may not be known to the analyst fitting the model because of confidentiality restrictions or for other reasons. Even when the design variables are known for the population, the model can be too complicated to be identified from the observed data.

More practical is to include all selection probabilities in the model when they are known for the entire population, which is rarely the case. Fortunately for some small area models it suffices to know population totals for selection probabilities.

**Nested error regression model with selection probabilities as additional covariates (Verret, Rao and Hidiroglou 2015):** Assume that all areas are selected and the selection within the areas is informative. In addition, assume that the nested error regression model with selection probabilities as additional covariates,

$$\begin{aligned} y_{ij} &= \tilde{\mathbf{x}}'_{ij}\boldsymbol{\beta} + u_i + e_{ij}; \quad u_i|I_i = 1 \overset{iid}{\sim} N(0, \sigma_u^2), \\ e_{ij}|I_{ij} &= 1 \overset{iid}{\sim} N(0, \sigma_e^2), \quad \tilde{\mathbf{x}}'_{ij} = [\mathbf{x}'_{ij}, g(\pi_{j|i})], \end{aligned} \tag{7.24}$$

for some function $g$, holds for the sample. If $g$ satisfies $E_{si}(\pi_{j|i}|\mathbf{x}_{ij}, g(\pi_{j|i}), I_i = 1) = \pi_{j|i} = E_U[\pi_{j|i}|\mathbf{x}_{ij}, g(\pi_{j|i})]$, for example $g(\pi_{j|i}) = \pi_{j|i}$, $g(\pi_{j|i}) = \log(\pi_{j|i})$, $g(\pi_{j|i}) = 1/\pi_{j|i}$ or $g(\pi_{j|i}) = n_i/\pi_{j|i}$, then by (7.5) and (7.7), the distribution of $e_{ij}$ in the population and the sample is the same, which in turn implies that the models are the same for the population and the sample, that is,

$$\begin{aligned} y_{ij} &= \tilde{\mathbf{x}}'_{ij}\boldsymbol{\beta} + u_i + e_{ij}; \quad u_i \overset{iid}{\sim} N(0, \sigma_u^2), \\ e_{ij} &\overset{iid}{\sim} N(0, \sigma_e^2), \quad \tilde{\mathbf{x}}'_{ij} = [\mathbf{x}'_{ij}, g(\pi_{j|i})]. \end{aligned} \tag{7.25}$$

Therefore in this case, similarly to (7.20),

$$\begin{aligned} \hat{\hat{Y}}_i &= \frac{1}{N_i}[\sum_{j\in s_i} y_{ij} + \sum_{l\notin s_i}(\tilde{\mathbf{x}}'_{il}\hat{\boldsymbol{\beta}} + \hat{u}_i)] \\ &= \frac{1}{N_i}\{\sum_{j\in s_i} y_{ij} + [\sum_{l\notin s_i}\tilde{\mathbf{x}}'_{il}]\hat{\boldsymbol{\beta}} + (N_i - n_i)\hat{u}_i\} \\ &= \frac{1}{N_i}\{\sum_{j\in s_i} y_{ij} + [\sum_{l=1}^{N_i}\tilde{\mathbf{x}}'_{il} - \sum_{l\in s_i}\tilde{\mathbf{x}}'_{il}]\hat{\boldsymbol{\beta}} + (N_i - n_i)\hat{u}_i\} \end{aligned} \tag{7.26}$$

is the empirical best predictor. This predictor requires the knowledge of population totals of $g(\pi_{j|i})$, which for some $g$ is available. For example, if the samples within the areas are of fixed sizes $n_i$, then $\sum_{j=1}^{N_i}\pi_{j|i} = n_i$ (case of $g(\pi_{j|i}) = \pi_{j|i}$). In addition, for some surveys the totals of $g(\pi_{j|i}) = n_i/\pi_{j|i}$ are also known. Verret, Rao and Hidiroglou (2015) show how $g$ can be fitted from the sample and test different choices of $g$ in simulation studies. An advantage of this predictor is that its MSE can be estimated using classical small area estimation theory (Rao and Molina 2015).

### 7.4.3 Informative Selection of Areas and Within the Areas, Prediction for a Sampled Area

Now, consider the general case when both, the selection of areas and the selection within the areas, are informative. We follow Pfeffermann and Sverchkov (2007). Their approach requires specifying the two-level sample model (7.17) and the conditional sample expectation, $E_{si}(w_{j|i}|\mathbf{x}_{ij}, y_{ij}, u_i, I_i = 1)$, all of which can be identified and tested using the observed data since they refer to sample models. We do not assume any model for the population data or unobserved data, and make no assumptions regarding the selection of areas, or the model holding for the area sample weights.

First, consider a sampled area. By (7.10), computation of the predictors of the area means requires estimating $E_s[E_{ci}(y_{il}|\mathbf{x}_{il}, u_i, I_i = 1)|D_s]$. By (7.7),

$$E_s[E_{ci}(y_{il}|\mathbf{x}_{il}, u_i, I_i = 1)|D_s] = E_s[H_{\mathbf{x}_{il}}(u_i)|D_s], \tag{7.27}$$

where

$$\begin{aligned} H_{\mathbf{x}_{il}}(u_i) &= E_{ci}(y_{il}|\mathbf{x}_{il}, u_i, I_i = 1) \\ &= \int y_{il} \frac{E_{si}(w_{l|i}|y_{il}, \mathbf{x}_{il}, u_i, I_i = 1) - 1}{E_{si}(w_{l|i}|, \mathbf{x}_{il}, u_i, I_i = 1) - 1} f_{si}(y_{il}|\mathbf{x}_{il}, u_i, I_i = 1) dy_{il}. \end{aligned}$$

If $y$ is discrete then the integral has to be replaced by a sum.

The integral $H_{\mathbf{x}_{il}}(u_i)$ depends on the sample model $f_{si}(y_{il}|\mathbf{x}_{il}, u_i, I_i = 1)$ and the expectation $E_{si}(w_{l|i}|y_{il}, \mathbf{x}_{il}, u_i, I_i = 1)$, which can be identified and estimated from the sample data, as can be seen from the examples which follow. The integral can be computed either analytically or, if necessary, using numerical approximations. The expectation $E_s[H_{\mathbf{x}_{il}}(u_i)|D_s]$ in (7.27) is with respect to the sample distribution $f_s(u_i|D_s, I_i = 1)$, which is obtained from the sample model defined by (7.1) and (7.3). This allows us, in principle, to compute $E_s[H_{\mathbf{x}_{il}}(u_i)|D_s]$, with unknown parameters replaced by sample estimates.

Alternatively, $E_s[H_{\mathbf{x}_{il}}(u_i)|D_s]$ can be estimated as,

$$\hat{E}_s[H_{\mathbf{x}_{il}}(u_i)|D_s] = \sum_{k \in s} H_{\mathbf{x}_{il}}(\hat{u}_k)/m, \tag{7.28}$$

where $\hat{u}_k$ is the predictor of $u_k$ given $D_s$ under the sample model. Finally, by (7.10), we obtain the predictor of the $i$'th area average,

$$\hat{\bar{Y}}_i = \frac{1}{N_i}\{\sum_{j \in s_i} y_{ij} + \sum_{l \notin s_i} \hat{E}_s[H_{\mathbf{x}_{il}}(u_i)|D_s]\}.$$

**Nested error regression model, prediction for a sampled area:** Suppose that the sampling weights, $w_{j|i}$ within the selected areas satisfy,

$$\begin{aligned} E_{si}(w_{j|i}|\mathbf{x}_{ij}, y_{ij}, u_i, I_i = 1) &= E_{si}(w_{j|i}|\mathbf{x}_{ij}, y_{ij}, I_i = 1) \\ &= k_i \exp(\mathbf{x'}_{ij}\mathbf{a} + by_{ij}), \end{aligned} \qquad (7.29)$$

where $k_i = N_i n_i^{-1} \sum_{j=1}^{N_i} \exp(\mathbf{x'}_{ij}\mathbf{a} + by_{ij})/N_i$, and $\mathbf{a}$ and $b$ are fixed (unknown) constants. If $x_{i0}$ is a constant, we assume $a_0 = 1$ for uniqueness. The form of $k_i$ follows from (7.6) and the well known fact from sampling theory that if $n_i$ units are selected by probability sampling from an area of size $N_i$ then $\sum_{j=1}^{N_i} \pi_{j|i} = n_i$. Note that for large areas by the law of large numbers,

$$\sum_{j=1}^{N_i} \exp(\mathbf{x'}_{ij}\mathbf{a} + by_{ij})/N_i \cong E_U[\sum_{j=1}^{N_i} \exp(\mathbf{x'}_{ij}\mathbf{a} + by_{ij})/N_i] = constant,$$

so that $k_i \cong (N_i/n_i) \times constant$. As will become evident below, for sufficiently small sampling fractions the predictors for sampled and non-sampled areas do not depend on $\mathbf{a}$ and $k_i$.

**Remark.** As with the sample model (7.18), the expectation in (7.29) refers to the sample distribution within the sampled areas. The relationship in the sample between the sampling weights and the outcomes can be identified and estimated therefore from the sample data using classical regression, under the assumption that the expectation in (7.29) does not depend on the random effects. Actually, (7.29) partly accounts for the random effect through $k_i$. On the other hand, the relationship between the sampling weights $w_i$ and the area means is more difficult to identify since the area means are not observable, and we do not model this relationship.

The analysis that follows assumes known model parameters for both, the sample model (7.18) and the selection model (7.29). In practice, the unknown model parameters are replaced under the frequentist approach by sample estimates, yielding the corresponding 'empirical predictors'. Maximum likelihood estimation of the model parameters has to be based in the present case on the sample distribution of the sample outcomes and weights. Alternatively, the model parameters can possibly be estimated by the method of moments, depending on the underlying model.

By (7.7), (7.18) and (7.29),

$$\begin{aligned} &f_{ci}(y_{il}|\mathbf{x}_{il}, u_i, I_i = 1) \\ &= \frac{[E_{si}(w_{l|i}|\mathbf{x}_{il}, y_{il}, u_i, I_i = 1) - 1] f_{si}(y_{il}|\mathbf{x}_{il}, u_i, I_i = 1)}{E_{si}(w_{l|i}|\mathbf{x}_{il}, u_i, I_i = 1) - 1} \\ &= \frac{\lambda_{il}}{\lambda_{il} - 1} \frac{1}{\sigma_e} \phi(\frac{y_{il} - u_{il} - b\sigma_e^2}{\sigma_e^2}) - \frac{\lambda_{il}}{\lambda_{il} - 1} \frac{1}{\sigma_e} \phi(\frac{y_{il} - u_{il}}{\sigma_e^2}), \end{aligned} \tag{7.30}$$

where $u_{il} = \mathbf{x}'_{il}\boldsymbol{\beta} + u_i$,

$$\lambda_{il} = k_i \exp[(b^2\sigma_e^2/2) + \mathbf{x}'_{il}\mathbf{a} + bu_{il}] = E_{si}(w_{l|i}|\mathbf{x}_{il}, u_i, I_i = 1),$$

and $\phi$ is the standard normal pdf. In the special case where $b = 0$ (the selection probabilities within the selected areas only depend on $\mathbf{x}$-values so that the sample is noninformative), the pdf in (7.30) reduces to the sample normal density (7.18). By computing the expectation under the sample-complement pdf (7.30) we find,

$$\begin{aligned} E_s[E_{ci}(y_{il}|\mathbf{x}_{il}, u_i, I_i = 1)|D_s] &= E_s[(u_{il} + \frac{\lambda_{il}}{\lambda_{il} - 1} b\sigma_e^2)|D_s] \\ &= (\mathbf{x}_{il}\boldsymbol{\beta} + \hat{u}_i) + b\sigma_e^2 E_s[(1 - \lambda_{il}^{-1})^{-1}|D_s], \end{aligned} \tag{7.31}$$

where the last equality follows from (7.19) and $\hat{u}_i = \gamma_i[\bar{y}_i - \mathbf{x}'_i\boldsymbol{\beta}]$.

The expectation $E_s[(1 - \lambda_{il}^{-1})^{-1}|D_s]$ can be computed numerically. Alternatively, in practical cases where the sampling fractions within the areas $n_i/N_i$ are small, $\lambda_{il} = E_s(w_{l|i}|\mathbf{x}_{il}, u_i, I_i = 1)$ is typically much larger than 1, and hence we may approximate $E_s[(1 - \lambda_{il}^{-1})^{-1}|D_s] \cong 1$, in which case by (7.31),

$$E_s[E_{ci}(y_{il}|\mathbf{x}_{il}, u_i, I_i = 1)|D_s] \cong (\mathbf{x}_{il}\boldsymbol{\beta} + \hat{u}_i) + b\sigma_e^2. \tag{7.32}$$

It follows from (7.10), (7.31) and (7.32) that for given estimates of $\{\boldsymbol{\beta}, b, \sigma_u^2, \sigma_e^2\}$, the mean $\bar{Y}_i$ can be predicted as,

$$\begin{aligned} \hat{\bar{Y}}_i &= \frac{1}{N_i}[\sum_{j\in s_i} y_{ij} + \sum_{l\notin s_i}(\mathbf{x}'_{il}\hat{\boldsymbol{\beta}} + \hat{u}_i)] + (N_i - n_i)\hat{b}\hat{\sigma}_i^2 \\ &= \hat{\bar{Y}}_{i,IGN} + (N_i - n_i)\hat{b}\hat{\sigma}_i^2, \end{aligned} \tag{7.33}$$

where $\hat{\bar{Y}}_{i,IGN}$ is the empirical best predictor of the sample model mean. That is, the empirical best predictor assuming noninformative selection of areas and within the areas, (7.20). The last term in (7.33) corrects for the sample selection effects, that is, the difference between the sample-complement expectation and the sample expectation in the sampled areas. For $b = 0$, the predictor (7.33) reduces to the optimal predictor, under noninformative sample, (7.20). Note that the predictor (7.33) does not depend on $k_i$ and $\mathbf{a}$ featured in the expectation (7.29).

**Logistic mixed model, prediction for a sampled area:** Suppose the sampling weights, $w_{j|i}$ within the selected areas satisfy,

$$\begin{aligned} E_{si}(w_{j|i}|\mathbf{x}_{ij}, y_{ij}, u_i, I_i = 1) &= E_{si}(w_{j|i}|\mathbf{x}_{ij}, y_{ij}, I_i = 1) \\ &= w_{\boldsymbol{\gamma}}^{(i)}(\mathbf{x}'_{ij}, y_{ij}), \end{aligned} \tag{7.34}$$

for some function depending on a parameter $\boldsymbol{\gamma}$. For example,

$$w_{\boldsymbol{\gamma}}^{(i)}(\mathbf{x}'_{ij}, y_{ij}) = c_i w_{\boldsymbol{\gamma}}(\mathbf{x}'_{ij}, y_{ij}), \quad w_{\boldsymbol{\gamma}}(\mathbf{x}'_{ij}, y_{ij}) = \frac{\exp(\mathbf{x}'_{ij}\boldsymbol{\gamma} + \gamma_y y_{ij})}{1 + \exp(\mathbf{x}'_{ij}\boldsymbol{\gamma} + \gamma_y y_{ij})}$$

is logistics, and similarly to $k_i$ in (7.29), $c_i = N_i n_i^{-1} \sum_{j=1}^{N_i} w_{\boldsymbol{\gamma}}(\mathbf{x}'_{ij}, y_{ij})/N_i$.

Suppose the logistic mixed model (7.22) is fitted to the sample and suppose that all parameters (or their sample estimates) in (7.22) and (7.34), and the predictors of the random effects $u_i$ in (7.22) are known. Denote $p(\mathbf{x}_{ij}, u_i) = Pr(y_{ij} = 1|\mathbf{x}_{ij}, u_i, I_{ij} = 1)$. Then $H_{\mathbf{x}_{il}}(u_i)$ in (7.27) can be written as,

$$H_{\mathbf{x}_{il}}(u_i) = \frac{w_{\boldsymbol{\gamma}}^{(i)}(\mathbf{x}'_{ij}, 1) - 1}{w_{\boldsymbol{\gamma}}^{(i)}(\mathbf{x}'_{ij}, 0)[1 - p(\mathbf{x}_{ij}, u_i)] + w_{\boldsymbol{\gamma}}^{(i)}(\mathbf{x}'_{ij}, 1)p(\mathbf{x}_{ij}, u_i) - 1} p(\mathbf{x}_{ij}, u_i),$$

and by (7.28),

$$\hat{\bar{Y}}_{i,IGN} = \frac{1}{N_i}[\sum_{j \in s_i} y_{ij} + \sum_{l \notin s_i}[\sum_{k \in s} \hat{H}_{\mathbf{x}_{il}}(\hat{u}_k)/m], \tag{7.35}$$

with $\hat{\boldsymbol{\gamma}}$ substituted for $\boldsymbol{\gamma}$ in $H_{\mathbf{x}_{il}}$.

### 7.4.4 Informative Selection of Areas and Within the Areas, Prediction for a Nonsampled Area

Consider nonsampled areas. By (7.11), predicting the means in such areas requires estimating, $E_c[E_U(y_{ik}|\mathbf{x}_{ik}, u_i, I_i = 1)|D_s]$. By (7.6) and the iterated expectation rule,

$$E_U(y_{ik}|\mathbf{x}_{ik},u_i,I_i=1)=\frac{E_{si}(w_{k|i}y_{ik}|\mathbf{x}_{ik},u_i,I_i=1)}{E_{si}(w_{k|i}|\mathbf{x}_{ik},u_i,I_i=1)}$$

$$=\frac{E_{si}[E_{si}(w_{k|i}|y_{ik},\mathbf{x}_{ik},u_i,I_i=1)y_{ik}|\mathbf{x}_{ik},u_i,I_i=1]}{E_{si}[E_{si}(w_{k|i}|y_{ik},\mathbf{x}_{ik},u_i,I_i=1)|\mathbf{x}_{ik},u_i,I_i=1]}$$

$$=\frac{\int E_{si}(w_{k|i}|y_{ik},\mathbf{x}_{ik},u_i,I_i=1)y_{ik}f_{si}(y_{ik}|\mathbf{x}_{ik},u_i,I_i=1)}{\int E_{si}(w_{k|i}|y_{ik},\mathbf{x}_{ik},u_i,I_i=1)f_{si}(y_{ik}|\mathbf{x}_{ik},u_i,I_i=1}=K_{\mathbf{x}_{ik}}(u_i).$$

$K_{\mathbf{x}_{ik}}(u_i)$ again only depends on the sample models $f_{si}(y_{ik}|\mathbf{x}_{ik},u_i,I_i=1)$ and $E_{si}(w_{k|i}|y_{ik},\mathbf{x}_{ik},u_i,I_i=1)$, and it can be computed analytically (see below) or numerically if necessary. Finally, by (7.8),

$$E_c[E_U(y_{ik}|\mathbf{x}_{ik},u_i,I_i=1)|D_s]=\frac{E_s[(w_i-1)K_{\mathbf{x}_{ik}}(u_i)|D_s]}{E_s[(w_i-1)|D_s]}=K_{\mathbf{x}_{ik}}(D_s)\text{(7.36)}$$

and the expectations in (7.36) can be estimated by the sample means,

$$\hat{K}_{\mathbf{x}_{ik}}(D_s)=\frac{\sum_{r\in s}(w_r-1)K_{\mathbf{x}_{ik}}(\hat{u}_r)}{\sum_{r\in s}(w_r-1)}. \tag{7.37}$$

By (7.11), the final predictor for the nonsampled areas is

$$\hat{\hat{Y}}_i=\sum_{k=1}^{N_i}\hat{K}_{\mathbf{x}_{ik}}(D_s)/N_i. \tag{7.38}$$

**Nested error regression model, prediction for a nonsampled area:** Computing the expectations

$$\int E_{si}(w_{k|i}|y_{ik},\mathbf{x}_{ik},u_i,I_i=1)y_{ik}f_{si}(y_{ik}|\mathbf{x}_{ik},u_i,I_i=1)$$

and

$$\int E_{si}(w_{k|i}|y_{ik},\mathbf{x}_{ik},u_i,I_i=1)f_{si}(y_{ik}|\mathbf{x}_{ik},u_i,I_i=1),$$

using (7.18) and (7.29) yields after some algebra,

$$E_c[E_U(y_{ik}|\mathbf{x}_{ik},u_i,I_i=1)|D_s]=\mathbf{x}'_{ik}\boldsymbol{\beta}+b\sigma_e^2+E[\frac{(w_i-1)u_i}{E_s(w_i|D_s)-1}|D_s]. \tag{7.39}$$

Estimating the two sample expectations on the right hand side of (7.39) by the corresponding sample means and substituting $\hat{u}_i = \gamma_i[\bar{y}_i - \bar{\mathbf{x}}'_i\boldsymbol{\beta}]$ for $u_i$ yields the following estimator for $E_{ik} = E_c[E_U(y_{ik}|\mathbf{x}_{ik}, u_i, I_i = 1)|D_s]$,

$$E_{ik} = \mathbf{x}'_{ik}\boldsymbol{\beta} + b\sigma_e^2 + \frac{\sum_{i\in s}(w_i - 1)\hat{u}_i}{\sum_{i\in s}(w_i - 1)}. \tag{7.40}$$

It follows from (7.11) and (7.40) that for given estimates of $\{\boldsymbol{\beta}, b, \sigma_u^2, \sigma_e^2\}$, the mean $\bar{Y}_i$ of area $i$, not in the sample, can be predicted as,

$$\hat{\bar{Y}}_i = \frac{1}{N_i}\sum_{k=1}^{N_i}\mathbf{x}'_{ik}\hat{\boldsymbol{\beta}} + \hat{b}\hat{\sigma}_i^2 + \frac{\sum_{i\in s}(w_i - 1)\hat{u}_i}{\sum_{i\in s}(w_i - 1)}. \tag{7.41}$$

The last term of (7.41) corrects for the fact that the mean of the random effects for areas outside the sample is different from zero under informative selection of areas.

**Logistic mixed model, prediction for a nonsampled area:** Under the models (7.34) and (7.22), similarly to the derivation of (7.35)

$$K_{\mathbf{x}_{ik}}(u_i) = \frac{E_{si}[E_{si}(w_{k|i}|y_{ik}, \mathbf{x}_{ik}, u_i, I_i = 1)y_{ik}|\mathbf{x}_{ik}, u_i, I_i = 1]}{E_{si}[E_{si}(w_{k|i}|y_{ik}, \mathbf{x}_{ik}, u_i, I_i = 1)|\mathbf{x}_{ik}, u_i, I_i = 1]}$$

$$= \frac{w_{\boldsymbol{\gamma}}^{(i)}(\mathbf{x}'_{ij}, 1)}{w_{\boldsymbol{\gamma}}^{(i)}(\mathbf{x}'_{ij}, 0)[1 - p(\mathbf{x}_{ij}, u_i)] + w_{\boldsymbol{\gamma}}^{(i)}(\mathbf{x}'_{ij}, 1)p(\mathbf{x}_{ij}, u_i)}p(\mathbf{x}_{ij}, u_i),$$

and the final predictor is defined by (7.38)

$$\hat{\bar{Y}}_i = \sum_{k=1}^{N_i}\frac{\sum_{r\in s}(w_r - 1)K_{\mathbf{x}_{ik}}(\hat{u}_i)}{\sum_{r\in s}(w_r - 1)}$$

with $\hat{\gamma}$ substituted for $\gamma$ in $K_{\mathbf{x}_{il}}$.

**Remark**. The MSE of the above predictors can be estimated by a parametric bootstrap (or double-bootstrap) procedure. Pfeffermann and Sverchkov (2007) illustrate the bootstrap procedure for the case when the model for the observed data is the nested error regression model and the selection model is (7.29). Note that such procedures are much more complicated than the classical ones assuming ignorable samples of areas and within areas, and probably less exact. See Pfeffermann and Sverchkov (2007) for MSE estimation under informative sampling of areas and within areas.

## 7.5 Testing for Prediction Bias

Evidently, predicting the small area means under informative sampling is more complicated and possibly less stable than under noninformative sampling. Thus, it is important to test the informativeness of the sample selection and if found noninformative, use standard optimal procedures. In what follows we propose simple test statistics for testing whether ignoring the sample selection biases the predictors.

### 7.5.1 Testing whether Ignoring the Selection of Areas Biases the Predictors

By (7.14), the selection of areas does not bias the predictors used under noninformative selection if $Cov_s\{[\sum_{k=1}^{N_i} E_U(y_{ik}|\mathbf{x}_{ik}, u_i, I_{ik} = 1), w_i]|D_s\} = 0$. Note that by (7.3) $E_U(y_{ik}|\mathbf{x}_{ik}, u_i, I_{ik} = 1) = E_{si}(y_{ik}|\mathbf{x}_{ik}, u_i, I_i = 1)$ and informativeness of the selection of areas can be tested by the hypothesis,

$$H_0 : Cov_s(L(u_i), w_i|D_s) = 0$$

where $L(u_i) = \sum_{k=1}^{N_i} E_{si}(y_{ik}|\mathbf{x}_{ik}, u_i, I_i = 1)$. $L(u_i)$ is a function of the sample model (7.15), and (7.15) has to be estimated anyway, regardless of whether the sample is informative or not.

For testing $H_0$ we would ideally regress $w_i$ against $L(u_i)$, but the random effects are unobservable and the parameters $\boldsymbol{\theta}$ of (7.15) are unknown. Thus, we regress instead $w_i$ against the estimates $L_{\hat{\boldsymbol{\theta}}}(\hat{u}_i)$ as computed under the sample model (7.15). Writing $L_{\hat{\boldsymbol{\theta}}}(\hat{u}_i) = L(u_i) + \eta_i$, it can be safely assumed that $Cov(w_i, \eta_i) = 0$, such that testing $H_0$ can be implemented by regressing $w_i = \delta_0 + \delta\hat{u}_i + \psi_i$ and testing $H_0 : \delta = 0$.

Assuming that $\psi_i$ are iid normal deviates, one can test $H_0$ using the $t$-statistics,

$$t^A = \frac{\hat{\delta}_{OLS}}{\sqrt{\hat{Var}(\hat{\delta}_{OLS})}} \tag{7.42}$$

Which has then a $t$-distribution with $(m - 2)$ degrees of freedom (holding the estimates $\hat{u}_i$ and $\hat{\boldsymbol{\theta}}$ fixed). The null hypothesis refers to the sample distribution, thus justifying estimating $\delta$ by ordinary least squares (OLS). For a large number of sampled areas (large $m$), the statistics $t^A$ retains approximately the $t$-distribution (effectively the standard normal distribution) even without the normality assumption, since the estimator $\hat{\delta}_{OLS}$ has asymptotically a normal distribution based on the central limit theorem. We mention also that regressing $w_i$ against $L_{\hat{\boldsymbol{\theta}}}(\hat{u}_i)$ and testing $H_0 : \delta = 0$ may not be very powerful if $Var(\eta_i)$ is large. Any alternative test could possibly be constructed by noting that $L_{\hat{\boldsymbol{\theta}}}(\hat{u}_i) = L(u_i) + \eta_i$, and using 'errors in variables techniques'.

### 7.5.2 Testing whether Ignoring the Sampling within the Areas Biases the Predictors

By (7.12), sampling within the areas does not bias the predictors used under noninformative sampling if $cov_{si}(y_{il}, w_{il}|\mathbf{x}_{ij}, u_i, I_i = 1) = 0$. Thus the ignorability of the sample selection within the selected areas can be tested by regressing $w_{l|i} = \gamma_{0i} + \mathbf{x}'_{il}\gamma_{1i} + \gamma_{2i}y_{il} + \eta_{il}$ and testing $H_0 : \gamma_{2i} = 0$, separately for every area $i$. (See the following remark). However, with a large number of sampled areas, testing $H_0$ for every area is not practical, and with small sample sizes within the selected areas the tests have low power. Assuming the same sampling design within the areas, a more practical and powerful test is therefore,

$$F^w_{max} = max\{F_i, i = 1, ..., m\}, \tag{7.43}$$

where $F_i$ defines the test statistics in area $i$. For a given distribution of $F_i$, computation of the percentiles of $F^w_{max}$ is straightforward. Here again, if the disturbances $\eta_{ij}$ can be assumed to be iid normal deviates, one can use the test statistics $F_i = [\hat{\gamma}_{2i}/\hat{SD}(\hat{\gamma}_{2i})]^2$, where $\hat{\gamma}_{2i}$ and $\hat{SD}(\hat{\gamma}_{2i})$ are, respectively, the OLS estimator and its estimated standard deviation. Under the null hypothesis $H_0 : \gamma_{2i} = 0$, $F_i \sim F(1, n_i - p - 2)$ where $p = dim(\mathbf{x}_{il})$. On the other hand, if the iid normality assumption is not warranted, the $F$ distribution cannot be justified by asymptotic arguments as in the case of the statistics $t^A$ in Section 7.5.1 since the sample sizes within the areas are typically small, and one has to use in this case more robust procedures.

**Remark.** Instead of testing $corr_{si}(y_{il}, w_{il}|\mathbf{x}_{ij}, u_i, I_i = 1) = 0$ by fitting a linear model, one can test

$$H^*_0 : E_{si}(w_{j|i}|y_{ij}, \mathbf{x}_{ij}, u_i, I_i = 1) = E_{si}(w_{j|i}|\mathbf{x}_{ij}, u_i, I_i = 1)$$

using other, more plausible relationships between the weights $w_{j|i}$ and $(y_{ij}, \mathbf{x}_{ij})$, such as (7.29). Note that under $H^*_0$ the marginal and sample distributions within the selected areas are the same.

## 7.6 Estimating Mean Body Mass Index in USA Counties

Pfeffermann and Sverchkov (2007) apply the methodology described in the previous sections in estimating the mean body mass index (BMI) for counties in the U.S. The BMI is defined as the ratio of the weight, measured in kilograms, and the square of the height, measured in meters. An index higher than 27.8 for men and higher than 27.3 for women defines overweight, which is known to be a major health risk factor. Estimating the mean BMI at the national and sub-national level is therefore of prime importance for public health authorities dealing with this problem. The data used for this study were collected as part of the third national health and nutrition examination survey (NHANES III). The survey was conducted in two phases during the years 1988-1991 and 1991-1994, and it represents the U.S. total civilian non-institutional population.

NHANES III is a stratified four-stage clustered survey that collects health, dietary and background information through questionnaires and physical examinations. The primary sampling units (PSU) are in most cases individual counties. There are 81 PSUs in the sample, selected with probability proportional to a measure of size without replacement. The size measure was constructed in such a way that the survey oversampled PSUs with large populations of Mexican-Americans and Blacks. The second stage of the sample selection consisted of sampling of area segments, which were then stratified based on the percent of Mexican-Americans. Next, households were sampled within the strata, with higher rates for strata with high minority concentrations. In the last stage a sample of persons was sampled from classes of households defined by age, sex and race that were sampled at different rates. For more details and the computation of the sampling weights, see `http://www.cdc.gov/nchs/about/major/nhanes/nh3data.htm`. The data set used for this study refers to the 81 sampled counties. There are 3138 counties in total in the U.S. The sample sizes within the sampled counties exceed 80 in almost all the counties, with a total sample size of 16,521, consisting of 8767 women and 7754 men. Thus, the major small area estimation problem associated with this survey is that only a small fraction of the counties that define the areas is represented in the sample.

In a previous article, Malec et al. (1999) used NHANES III data for estimating overweight prevalence for states in the U.S. by fitting logistic models with fixed age/race/gender effects and random race/gender effects. In order to account for sampling effects within the selected counties, the authors estimated the sampling probabilities utilizing the sampling weights, and then substituted the estimates in the likelihood. The state prevalence estimates were obtained by applying the Bayesian approach with the resulting empirical likelihood, using MCMC simulations.

Pfeffermann and Sverchkov (2007) fit the model (7.18), separately for men and women, with county random effects and seven covariates: A constant, 3 dummy race variables, and 3 age variables. The race variables are: $x_1 = 1$ if non Hispanic white, $x_2 = 1$ if non Hispanic black, and $x_3 = 1$ if Hispanic. The age variables are: $age \times I_{20<age<50}$, $age \times I_{50 \leq age<75}$, $age \times I_{75 \leq age}$. The age variables are used as proxies for a quadratic relationship between the BMI and age (the county means of $age^2$ are unknown and that is why this variable cannot be included in model 7.18). There are a few other covariates with sample measurements that affect the BMI but could not be used for the same reason. One of these variables is education, measured by the number of school years, which was found to have a negative effect on the BMI of women. The data files that could be used contain information only on the county numbers of adults with college and higher education, but this information is unknown on the individual level.

Table 7.1: Regression coefficients and SE in parentheses. Variance estimates for the BMI Models: (Men) $\hat{\sigma}_u^2 = .760$, $\hat{\sigma}_e^2 = 23.040$; (Women) $\hat{\sigma}_u^2 = 2.830$, $\hat{\sigma}_e^2 = 39.560$

| Coefficient | Intercept | $x_1$ | $x_2$ | $x_3$ | $x_4$ | $x_5$ | $x_6$ |
|---|---|---|---|---|---|---|---|
| Men | 22.960<br>(0.414) | 0.739<br>(0.314) | 0.740<br>(0.316) | 1.161<br>(0.322) | 0.083<br>(0.008) | 0.056<br>(0.005) | 0.020<br>(0.004) |
| Women | 21.852<br>(0.526) | -0.670<br>(0.374) | 2.355<br>(0.375) | 1.602<br>(0.394) | 0.133<br>(0.010) | 0.095<br>(0.006) | 0.049<br>(0.005) |

Table 7.1 shows the estimated regression coefficients, their standard errors (SE), and (in the caption) the estimates of the variance of the random effects and the residual variance. All the coefficients except in the case of 'White non Hispanic' in the women's model are significant at the 5% level based on the conventional $t$-statistic. Pfefermann and Sverchkov (2007) tested the assumption that the residual variance is constant across the counties by first fitting the model for each of the sampled counties separately, assuming fixed county effects and then testing the homogeneity of the estimated residuals. After dropping 7 outlying counties, the hypothesis of homogeneity is accepted using Bartletts test with $p$-values of 0.99 for women, and 0.13 for men.

Next Pfefermann and Sverchkov (2007) applied the tests of sample ignorability considered in Section 7.5. They found that for both men and women the sampling within the counties does not introduce prediction bias (given the covariates included in the model), and that the sampling of counties is informative for women, but not for men. The $p$-values when testing the sample ignorability within the counties are 0.56 for women, and 0.41 for men. The sample ignorability within the counties has been tested also by regressing $\log(w_{j|i})$ against $\mathbf{x}_{ij}, y_{ij}$ instead of regresssing $w_{j|i}$, and by fitting the two regression models in each of the sampled counties separately, confirming in all the cases that for the present model the sample selection within the counties can be ignored. On the other hand, when testing the ignorability of the county selection using (7.42), the $p$-values are 0.0164 for women, and 0.31 for men, suggesting an informative sampling of counties for the women's model but not for the men's model.

As explained above, the sampling probabilities within the counties were determined by race and age characteristics, and hence it is not surprising that the sampling within the counties was found to be ignorable for the present model that includes race and age as explanatory variables. It is interesting to mention in this regard that Malec et al. (1999) found that the sampling within the counties is informative, despite the fact that their model likewise accounts for age and race/gender categories. The authors do not elaborate on the reasons for this finding but they show results illustrating different national and state estimates, depending on whether the sampling process is accounted for or not.

The result that the sampling of counties is informative for the women's model is likewise not surprising because the county selection probabilities were determined by the true county race totals, and these totals are not included in the model (see below). The model of Malec et al. (1999) contains fixed and random race parameters, which is probably why the authors concluded that the selection of counties is not informative for their model. The fact that the selection of counties can be ignored for the men's model in present application is probably related to the fact that the variance of the county random effects is small, $\hat{\sigma}_u^2 = 0.76$ , which makes it harder to detect selection effects.

As mentioned in Section 7.4.2, a possible way of controlling sampling effects is by including in the model all the design variables used for the sample selection. In the present application we are in a fortunate (but uncommon) situation where the county design variables: $x_{8i}$ = county total of non Hispanic White, $x_{9i}$ = county total of non Hispanic Black, and $x_{10i}$ = county total of Hispanic, are known. Adding these variables (divided by $10^5$) to the model yields the following coefficients and standard errors. Women:

$\beta_8 = -0.112(0.076)$, $\beta_9 = -0.089(0.200)$, $\beta_{10} = 0.141(0.141)$. Men: $\beta_8 = -0.017(0.043)$, $\beta_9 = -0.064(0.115)$, $\beta_{10} = 0.037(0.079)$. The coefficients and standard errors of the other covariates change only slightly from their values in Table 7.1 when fitting the model with only the six covariates, $x_1, ..., x_6$. Thus, all three design variables are highly insignificant individually, and they are also jointly insignificant with p-values of 0.42 for women and 0.69 for men. With such high p-values, many analysts would tend to drop the design variables from the model and conclude that the sampling of counties is noninformative for the six covariates model, which in view of the previous analysis is not true for the women's model. Furthermore, when re-estimating the random effects using the extended model that includes the three design variables, and applying the informativeness test in (7.42), Pfefermann and Sverchkov (2007) found that the sampling of counties is not informative for this model, with $p$-values 0.17 for women and 0.63 for men. Thus, the selection of counties can only be ignored when including the design variables in the model.

What are the implications of the use of the model with six covariates or the model with 9 covariates (including the 3 design variables) on the prediction of the county means? In what follows we restrict attention to the models for women because the selection of counties was found earlier to be ignorable for the men's model. Starting with the sampled areas, both models yield very similar predictions when using the predictors defined by (7.20), which are the empirical best linear predictors (EBLUP) under non-informative sampling within the areas. For the nonsampled areas, however, predictors defined by (7.21) yield somewhat different predictions. Figure 7.1 shows four different predictions of the means in nonsampled areas: The prediction based on (7.21) under the reduced model (6 covariates) as obtained when ignoring the county selection, the prediction based on (7.21) under the extended model with 9 regressors, (the vector contains in this case both the proportions and the totals of the three races), the prediction based on (7.41) under the reduced model ($b = 0$), and the prediction based on (7.41) under the extended model. The horizontal line at 27.3 marks the threshold defining overweight. For the predictor (7.41) under the reduced model the area selection bias correction,

$$\sum_{i \in s}(w_i - 1)\hat{u}_i / \sum_{i \in s}(w_i - 1)$$

is 0.47, with Jackknife estimated standard deviation of 0.16. For the predictor (7.41) under the extended model the bias correction is 0.25 with similar estimated standard deviation. The use of the bias correction for the

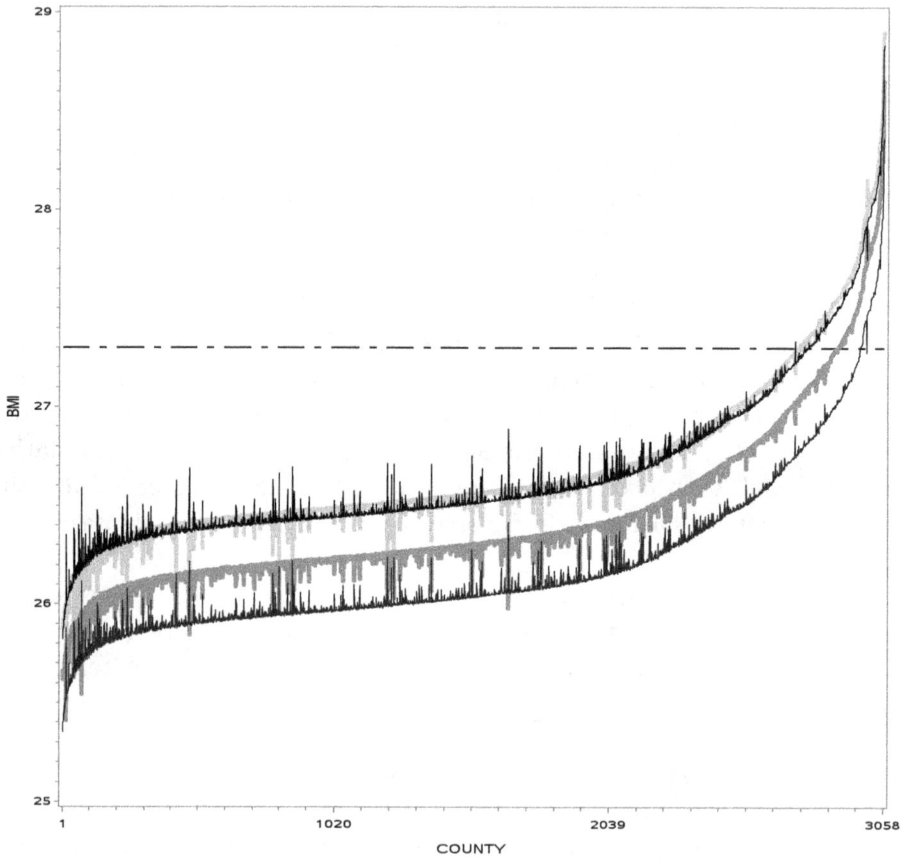


Figure 7.1: Prediction of mean body mass index of women in nonsampled counties of NHANES III. Values above the horizontal line at 27.3 define 'overweight'. The lower dark and grey curves show the synthetic predictors under the six covariates model and the 9 covariates model respectively. The upper dark and grey curves show the corresponding predictors with bias corrections. The counties are ordered by the average values of the 4 predictors.

extended model is therefore questionable, which is consistent with the test result that the selection of counties is ignorable for this model. The use of Jackknife for variance estimation assumes that the random effects $\hat{u}_i$ are approximately independent. It is used here only as a rough measure for assessing the stability of the bias correction.

The 4 curves in Figure 7.1 suggest that ignoring the county selection method and using the synthetic predictor based on the 6 regressors model under-predicts the true county means. This becomes evident by comparing the synthetic predictions under this model with the synthetic predictions under the extended model. The latter predictions are lower than the predictions obtained under the 6 covariates model with the bias correction, but interesting enough, once the bias correction is added also to the predictors under the extended model, both sets of predictions behave very similarly. However, as discussed above, the use of a bias correction for the extended model is questionable.

The magnitudes of the bias corrections seem very small, but they are not negligible. To see this, Pfeffermann and Sverchkov (2007) computed the percentage of nonsampled areas for which the predicted means are higher than the threshold of 27.3, as obtained by use of the four predictors. The use of the two synthetic predictors yields a percentage of 2.84% for the six covariates model and 5.56% for the extended model. Adding the bias correction of 0.47 to the first synthetic predictor increases the percentage to 9.2%, whereas adding the bias correction of 0.25 to the second synthetic predictor further increases the percentage to 10.3%. Thus, if areas with means that exceed the threshold are to be given extra attention, the use of the bias correction can be very important.

## 7.7 Appendix: Prediction of Small Area Means in the Presence of Informative Nonresponse

In the previous sections we assumed full response. But in reality almost all surveys are subject to nonresponse. In this section we assume that every unit $(i, j)$ selected by a two-stage selection process described in section 7.1 does not respond ($y_{ij}$ cannot be measured by some reason) with unknown probability, and denote by $R_{ij}$ a unit response indicator, $R_{ij} = 1$ if there is a unit response and 0 otherwise. Denote by $R = \{(i, j) : I_i = 1, I_{ij} = 1, R_{ij} = 1\}$ a set of respondents, and by $R^c = \{(i, l) : I_i = 1, I_{il} = 1, R_{il} = 0\}$ a set of nonrespondents, and let

$$O = \{y_{ij}, \pi_i, \pi_{j|i}, n_i, (i, j) \in R, \mathbf{x}_{kh}, k = 1, ..., M; h = 1, ..., N_k\}$$

be the observed data.

The response process is assumed to occur stochastically, independently between units. The observed sample of respondents can be viewed therefore as the result of a two-phase sampling process where in the first phase the sample is selected from $U$ with known inclusion probabilities $\pi_i, \pi_{j|i}$, and in the second phase the sample $R$ is 'self selected' with unknown response probabilities (Särndal and Swensson, 1987).

Denote

$$p(y_{ij}, \mathbf{x}_{ij}) = Pr[(i,j) \in R | y_{ij}, \mathbf{x}_{ij}, I_i = 1, I_{ij} = 1].$$

If $p(y_{ij}, \mathbf{x}_{ij})$ were known, the sample of respondents could be considered as a two-stage sample from a finite population with known selection probabilities $\pi_i$ and $\tilde{\pi}_{j|i} = \pi_{j|i} p(y_{ij}, \mathbf{x}_{ij})$, and the results of the previous sections could be applied to the case of non-response.

Also, if known, the response probabilities could be used for imputation within the selected areas via the relationship

$$\begin{aligned} &f(y_{il}|\mathbf{x}_{il}, u_i, (i,l) \in R^c) \\ &= \frac{[p^{-1}(y_{il}, \mathbf{x}_{il}) - 1] f(y_{il}|\mathbf{x}_{il}, u_i, (i,l) \in R)}{E_U\{[p^{-1}(y_{il}, \mathbf{x}_{il}) - 1] | \mathbf{x}_{il}, u_i, (i,l) \in R)\}}. \end{aligned} \tag{7.44}$$

Expression (7.44) follows directly from (7.7). Note that again the right hand side of (7.44) refers to the model for the observed data and therefore can be estimated by classical methods that ignore nonresponse and selection mechanisms.

In the rest of this section we concentrate on estimation of the response probabilities $p(y_{ij}, \mathbf{x}_{ij})$. If $p(y_{ij}, \mathbf{x}_{ij})$ does not depend on outcome $y_{ij}$, nonresponse missing at random, (MAR, Rubin, 1976; Little, 1982), $p(\mathbf{x}_{ij})$ can be estimated by classical regression techniques, for example, by assuming a parametric model

$$p(\mathbf{x}_{ij}; \gamma^{MAR}) = Pr[(i,j) \in R | y_{ij}, \mathbf{x}_{ij}, I_i = 1, I_{ij} = 1; \gamma^{MAR}],$$

and supposing that $p$ is differentiable with respect to the (vector) parameter $\gamma^{MAR}$. Note that this model refers to the observed data and therefore can be tested. Then $\gamma^{MAR}$ can be estimated, for example, by solving log-likelihood equations,

$$\sum_{(i,j) \in R} \frac{\partial \log[p(\mathbf{x}_{ij}; \gamma^{MAR})]}{\partial \gamma^{MAR}} + \sum_{(i,l) \in R^c} \frac{\partial \log[1 - p(\mathbf{x}_{il}; \gamma^{MAR})]}{\partial \gamma^{MAR}} = 0. \tag{7.45}$$

In many practical situations the MAR assumption is not valid, since the probability of responding often depends directly on the outcome value (response not missing at random, NMAR). In this case, the use of methods that assume that the nonresponse is MAR can lead to large biases of population parameter estimators and a large imputation bias.

Sverchkov (2008) suggested the following approach to estimating response probabilities under NMAR. Assume a parametric model

$$p(y_{ij}, \mathbf{x}_{ij}; \gamma) = Pr[(i,j) \in R | y_{ij}, \mathbf{x}_{ij}, I_i = 1, I_{ij} = 1; \gamma],$$

and suppose that $p$ is differentiable with respect to the (vector) parameter $\gamma$. If the missing data were observed, $\gamma$ could be estimated by solving the log-likelihood equations:

$$\sum_{(i,j)\in R} \frac{\partial \log[p(y_{ij}, \mathbf{x}_{ij}; \gamma)]}{\partial \gamma} + \sum_{(i,l)\in R^c} \frac{\partial \log[1 - p(y_{il}, \mathbf{x}_{il}; \gamma)]}{\partial \gamma} = 0. \quad (7.46)$$

Similarly to the Missing Information Principle (Cipillini et al, 1955, Orchard and Woodbury 1972), since the outcome values are missing for $(i,j) \in R^c$, Sverchkov (2008) suggested to solve instead of (7.46),

$$\begin{aligned} 0 &= E_U[\sum_{(i,j)\in R} \frac{\partial \log[p(y_{ij}, \mathbf{x}_{ij}; \gamma)]}{\partial \gamma} + \sum_{(i,l)\in R^c} \frac{\partial \log[1 - p(y_{il}, \mathbf{x}_{il}; \gamma)]}{\partial \gamma} | O] \\ &= \sum_{(i,j)\in R} \frac{\partial \log[p(y_{ij}, \mathbf{x}_{ij}; \gamma)]}{\partial \gamma} + \sum_{(i,l)\in R^c} E_U[\frac{\partial \log[1 - p(y_{il}, \mathbf{x}_{il}; \gamma)]}{\partial \gamma} | O] \\ &= \sum_{(i,j)\in R} \frac{\partial \log[p(y_{ij}, \mathbf{x}_{ij}; \gamma)]}{\partial \gamma} \\ &+ \sum_{(i,l)\in R^c} E_U\{E_U[\frac{\partial \log[1 - p(y_{il}, \mathbf{x}_{il}; \gamma)]}{\partial \gamma} | \mathbf{x}_{il}, u_i, (i,l) \in R^c] | O\}. \quad (7.47) \end{aligned}$$

In the last equality, similarly to the remark at Section 7.2, we assume for simplicity that $f(y_{il} | O, u_i, (i,l) \notin R) = f(y_{il} | \mathbf{x}_{il}, u_i, (i,l) \notin R)$. By (7.44),

$$\begin{aligned} & E_U[\frac{\partial \log[1 - p(y_{il}, \mathbf{x}_{il}; \gamma)]}{\partial \gamma} | \mathbf{x}_{il}, u_i, (i,l) \in R^c] \\ &= \int \frac{\partial \log[1 - p(y_{il}, \mathbf{x}_{il}; \gamma)]}{\partial \gamma} f(y_{il} | \mathbf{x}_{il}, u_i, (i,l) \notin R) dy_{ij} \\ &= \int \frac{[p^{-1}(y_{il}, \mathbf{x}_{il}) - 1] \frac{\partial \log[1 - p(y_{il}, \mathbf{x}_{il}; \gamma)]}{\partial \gamma} f(y_{il} | \mathbf{x}_{il}, u_i, (i,l) \in R)}{E_U\{[p^{-1}(y_{il}, \mathbf{x}_{il}) - 1] | \mathbf{x}_{il}, u_i, (i,l) \in R)\}} dy_{il}. \quad (7.48) \end{aligned}$$

(If $y$ is discrete then integrals have to be replaced by the sums.) Therefore response probabilities can be estimated by the following two stage procedure:

Step 1: Fit and estimate the observed model $f(y_{ij}|\mathbf{x}_{ij}, u_i, (i,j) \in R)$ as in section 7.4.1.

Step 2: Substitute the estimated model into (7.47) - (7.48) and solve the equations.

We note that Step 2 requires the integration of

$$E_U[\frac{\partial \log[1 - p(y_{il}, \mathbf{x}_{il}; \gamma)]}{\partial \gamma}|\mathbf{x}_{il}, u_i, (i,l) \in R^c]$$

over the distribution of $u_i|O$. Also,

$$E_U\{E_U[\frac{\partial \log[1 - p(y_{il}, \mathbf{x}_{il}; \gamma)]}{\partial \gamma}|\mathbf{x}_{il}, u_i, (i,l) \in R^c]|O\}$$

can be approximated by

$$E_U[\frac{\partial \log[1 - p(y_{il}, \mathbf{x}_{il}; \gamma)]}{\partial \gamma}|\mathbf{x}_{il}, \hat{u}_i, (i,l) \in R^c]$$

where, as above, $\hat{u}_i$ is a predictor of the random effect; see the following example.

**Logistic mixed model for outcomes:** Suppose logistic mixed model fits well the observed (after nonresponse) data,

$$Pr(y_{ij} = 1|\mathbf{x}_{ij}, u_i, I_{ij} = 1, R_{ij} = 1) = \frac{\exp(\mathbf{x'}_{ij}\boldsymbol{\beta} + u_i)}{1 + \exp(\mathbf{x'}_{ij}\boldsymbol{\beta} + u_i)},$$
$$u_i \overset{iid}{\sim} N(0, \sigma_u^2), \qquad (7.49)$$

and let $f_y(\mathbf{x'}_{ij}, u_i) = \frac{\exp(\mathbf{x'}_{ij}\hat{\boldsymbol{\beta}}+u_i)}{1+\exp(\mathbf{x'}_{ij}\hat{\boldsymbol{\beta}}+u_i)}$ be the estimate of (7.49). Let

$$p(y_{ij}, \mathbf{x}_{ij}; \gamma) = Pr[(i,j) \in R|y_{ij}, \mathbf{x}_{ij}, I_i = 1, I_{ij} = 1; \gamma]$$

be some parametric working model for the non-response. Then (7.48) can be written as

$$E_U[\frac{\partial \log[1 - p(y_{il}, \mathbf{x}_{il}; \gamma)]}{\partial \gamma}|\mathbf{x}_{il}, u_i, (i,l) \in R^c]$$

$$= \{[p^{-1}(0, \mathbf{x}_{il}) - 1]\frac{\partial \log[1 - p(0, \mathbf{x}_{il}; \gamma)]}{\partial \gamma}[1 - f_y(\mathbf{x'}_{il}, u_i)]$$

$$+ [p^{-1}(1, \mathbf{x}_{il}) - 1] \frac{\partial \log[1 - p(1, \mathbf{x}_{il}; \gamma)]}{\partial \gamma} f_y(\mathbf{x}'_{il}, u_i)\}$$

$$/\{[p^{-1}(1, \mathbf{x}_{il}) - 1] f_y(\mathbf{x}'_{il}, u_i) + [p^{-1}(0, \mathbf{x}_{il}) - 1][1 - f_y(\mathbf{x}'_{il}, u_i)]\},$$

and (7.47) can be approximated by

$$\begin{aligned} 0 &= \sum_{(i,j)\in R} \frac{\partial \log[p(y_{ij}, \mathbf{x}_{ij}; \gamma)]}{\partial \gamma} \\ &+ \sum_{(i,l)\notin R} E_U[\frac{\partial \log[1 - p(y_{il}, \mathbf{x}_{il}; \gamma)]}{\partial \gamma} | \mathbf{x}_{il}, \hat{u}_i, (i,l) \in R^c]. \end{aligned} \tag{7.50}$$

**Remark.** The main disadvantage of this approach is that the model for nonresponse can not be tested from the observed data and therefore can be easily misspecified. Another problem is identifiability of the parameters. Hence it is suggested to use simple models for the nonresponse; see Sverchkov (2013) for some discussion. Riddless, Kim and Im (2016) considered the estimation of response probabilities based on (7.47) and found some sufficient conditions for the identifiability of the solution of (7.47).

# References


1. Agresti, A. and Coull, B. A. (1998). Approximate is better than "exact" for interval estimation of binomial proportions. *American Statistician*, **52**, 119-126.

2. Aitchison, J. (1986). *Statistical Analysis of Compositional Data.* Chapman and Hall, New York.

3. Aitchison, J. and Shen, S.M. (1980). Logistic-normal distributions: Some properties and uses. *Biometrika*, **67**, 261-272.

4. Ahmad, I.A. (1995). On multivariate kernel estimation for samples from weighted distributions. *Statistics and Probability Letters*, **22**, 121-129.

5. Alizadeh, A.A., Eisen, M.B., Davis, R.E., Ma, C., Lossos, I.S., Rosenwald, A., Boldrick, J.C., Sabet, H., Tran, T., Yu, X., Powell, J.I., Yang, L., Marti, G.E., Moore, T., Hudson, J. Jr, Lu, L., Lewis, D.B., Tibshirani, R., Sherlock, G., Chan, W.C., Greiner, T.C., Weisenburger, D.D., Armitage, J.O., Warnke, R., Levy, R., Wilson, W., Grever, M.R., Byrd, J.C., Botstein, D., Brown, P.O., and Staudt, L.M., (2000). Distinct types of diffuse large B-cell lymphoma identified by gene expression profiling. *Nature*, **403**, 503-511.

6. Aldous, D. (1989). *Probability Approximations via the Poisson Clumping Heuristic.* Springer, New York.

7. Anderson, J.A. (1972). Separate sample logistic discrimination. *Biometrika*, **59**, 19-35.

8. Anderson, J.A. (1979). Multivariate logistic compounds. *Biometrika*, **66**, 17-26.

9. Anderson, T.W. (1971). *An Introduction to Multivariate Statistical Analysis.* Wiley, New York.

10. Ando, T. (2010). *Bayesian Model Selection and Statistical Modeling.* CRC Press, Boca Raton.

11. Arora, V., and Lahiri, P. (1997). On the superiority of the bayes method over the BLUP in small area estimation problems. *Statistica Sinica,* **7**, 1053-1063.

12. Battese, G.E., Harter, R.M., and Fuller, W.A. (1988). An error component model for prediction of county crop areas using survey and satellite data. *Journal of the American Statistical Association,* **83**, 28-36.

13. Bickel, P.J., Klaassen, C.A.J., Ritov, Y. and Wellner, J.A. (1998). *Efficient and Adaptive Estimation for Semiparametric Models.* Springer, New York.

14. Bondell, H.D. (2007). Testing goodness-of-fit in logistic case-control studies. *Biometrika,* **94**, 487-495.

15. Box, G.E.P., Jenkins, G.M., and Reinsel, G.C. (1994). *Time Series Analysis: Forecasting and Control,* 4th Ed. Wiley, Hoboken.

16. Brown, L.D., Cai, T.T. and DasGupta, A. (2001). Interval estimation for a binomial proportion. *Statistical Science,* **16**, 101-133.

17. Cai, S. (2014). *On dual empirical likelihood inference under semiparametric density ratio models in the presence of multiple samples.* Ph.D. dissertation, The University of British Columbia.

18. Cepillini, R., Siniscialco, M., and Smith, C.A.B. (1955). The estimation of gene frequencies in a random mating population. *Annals of Human Genetics,* **20**, 97-115.

19. Chamberlain, G. and Imbens, G.W. (2003). Nonparametric applications of Bayesian inference. *Journal of Business and Economic Statistics.* **21**, 12-18.

20. Chaudhuri, S. and Ghosh, M. (2011). Empirical likelihood for small area estimation. *Biometrika,* **98**, 473-480.

21. Chen, J. and Liu, Y. (2016). Small area estimation under density ratio model. `http://citeseerx.ist.psu.edu/viewdoc/download?doi=10.1.1.710.435&rep=rep1&type=pdf`

22. Chen, J. and Qin, J. (1993). Empirical likelihood estimation for finite populations and the effective usage of auxiliary information. *Biometrika*, **80**, 107-116.

23. Cheng, K.F. and Chu, C.K. (2004). Semiparametric density estimation under a two-sample density ratio model. *Bernoulli*, **10**, 583-604.

24. Cox, D.R. and Snell, E.J. (1989). *The Analysis of Binary Data* (2nd ed.). Chapman & Hall, London.

25. Crain, N.V, Crain, W.M., McQuillan, L.J. and Abramyan, H. (2009). Tort law tally: How state tort reforms affect tort losses and tort insurance premiums. Pacific Research Institute, San Francisco, CA.

26. Datta, G.S. (2009). Model-based Approaches to Small Area estimation. Chapter 32 of *Handbook of Statistics 29: Sample Surveys: Design, Methods and Applications* (Editors Pfeffermann, D. and Rao, C. R.). Amsterdam: North Holland.

27. Dayaratna, K. (2014). *Contributions to Bayesian statistical modeling in public policy research.* Ph.D. Dissertation, University of Maryland, College Park.

28. De Oliveira, V., Kedem, B. and Short, D.A. (1997). Bayesian prediction of transformed Gaussian random fields. *Journal of the American Statistical Association*, **92**, 1422-1433.

29. De Oliveira, V. and Kedem, B. (2017). Bayesian analysis of a density ratio model. *The Canadian Journal of Statistics.* To appear.

30. Devesa, S.S., Sigurdson, A.J., Brown, L.M., Tsao L. and Tarone R.E. (2003). Trends in the incidence of testicular germ cell tumors in the United States. *Cancer*, **97**, 63-70.

31. Fay, R.E., and Herriot, R.A. (1979). Estimation of income from small places: An application of James-Stein procedures to census data. *Journal of the American Statistical Association*, **74**, 269-277.

32. Fithian, W. and Wager, S. (2015). Semiparametric exponential families for heavy-tailed data. *Biometrika*, **102**: 486-493.

33. Fokianos, K. and Kaimi, I. (2006). On the effect of misspecifying the density ratio model. *Annals of the Institute of Statistical Mathematics*, **58**, 475-497.

34. Fokianos, K., Kedem, B., Qin, J. and Short, D.A. (2001). A semiparametric approach to the one-way layout. *Technometrics*, **43**, 56-65.

35. Fokianos, K. (2004). Merging information for semiparametric density estimation. *Journal of the Royal Statistical Society, Series B*, **66**, 941-958.

36. Fokianos, K. and Qin, J. (2008). A note on monte carlo maximization by the density ratio model. *Journal of Statistical Theory and Practice*, **2**, 355-367.

37. Gagnon R. (2005). *Computational aspects of power efficiency and state space models.* Ph.D. Dissertation, University of Maryland, College Park.

38. Gagnon, R., Kedem, B. and Qi, Y. (2008). On the efficiency of a semiparametric approach to the one-way layout. *Journal of Statistical Theory and Practice*, **2**, 385-406.

39. Gamerman, D. and Lopes, H.F. (2006). *Markov Chain Monte Carlo: Stochastic Simulation for Bayesian Inference*, 2nd. Ed. Chapman & Hall, Boca Raton.

40. Gill, R.D., Vardi, Y. and Wellner, J.A. (1988). Large sample theory of empirical distributions in biased sampling models. *Annals of Statistics*, **16**, 1069-1112.

41. Gilbert, P.B., Lele, S.R. and Vardi, Y. (1999). Maximum likelihood estimation in semiparametric selection bias models with application to AIDS vaccine trials. *Biometrika*, **86**, 27-43.

42. Gilbert, P.B. (2000). Large sample theory of maximum likelihood estimates in semiparametric biased sampling models. *Annals of Statistics*, **28**, 151-194.

43. Guo, H. (2005). *Generalized volatility model and calculating VaR using a new semiparametric model.* Ph.D. dissertation, University of Maryland, College Park.

44. Han, C. and Carlin, B.P. (2001). Markov chain Monte Carlo methods for computing Bayes factors: A comparative review. *Journal of the American Statistical Association*, **96**, 1122-1132.

45. Hall, P. and Weissman, I. (1997). On the estimation of extreme tail probabilities. *The Annals of Statistics*, **25**, 1311-1326.

46. Jones, M.C. (1991). Kernel density estimation for length bias data. *Biometrika*, **78**, 511-519.

47. Kass, R.E. and Raftery, A.E. (1995). Bayes factors. *Journal of the American Statistical Association*, **90**, 773-795

48. Katzoff, M., Zhou, W., Khan, D., Lu, G. and Kedem, B. (2014). Out-of-sample fusion in risk prediction. *Journal of Statistical Theory and Practice*, **8**, 444-459.

49. Kay, R., and S. Little (1987). Transformations of the explanatory variables in the logistic regression model for binary data. *Biometrika*, **74**, 495-501.

50. Kedem, B. and Fokianos, K. (2002). *Regression Models for Time Series Analysis.* Wiley, Hoboken.

51. Kedem, B., Gagnon, R.E. and Guo, H. (2005). Time Series prediction via density ratio modeling. Unpublished.

52. Kedem, B., Pan, L., Zhou, W., and Coelho, C.A. (2016). Interval estimation of small tail probabilities - Applications in food safety. *Statistics in Medicine*, **35**, 3229-3240.

53. Kedem, B., Wolff, D.B. and Fokianos, K. (2004). Statistical comparison of algorithms. *IEEE Transactions on Instrumentation and Measurement*, **53**, 770-776.

54. Kedem, B., Wen, S. (2007). Semi-parametric cluster detection. *Journal of Statistical Theory and Practice*, **1**, 49-72.

55. Kedem, B., Kim, E-Y, Voulgaraki, A. and Graubard, B. (2009). Two-dimensional semiparametric density ratio modeling of testicular germ cell data. *Statistics in Medicine*, **28**, 2147-2159.

56. Kedem, B. and Gagnon, R.E. (2010). Semiparametric distribution forecasting. *Journal of Statistical Planning and Inference*, **140**, 3734-3741.

57. Kedem, B., Lu, G., Wei, R. and Williams, D. (2008). Forecasting mortality rates via density ratio modeling. *Canadian Journal of Statistics*, **36**, 193-206.

58. Keziou, A. and Leoni-Aubin, S. (2008). On empirical likelihood for semiparametric two-sample density ratio models. *Journal of Statistical Planning and Inference*, **138**, 915-928

59. Kiefer, J. and Wolfowitz, J. (1956). Consistency of the maximum likelihood estimator in the presence of infinitely many incidental parameters. *Annals of Mathematical Statistics*, **27**, 887-906.

60. Kim, D.H. (2002), Bayesian and empirical bayesian analysis under informative sampling. *Sankhyā*, Ser. B, **64**, 267-288.

61. Kitamura, Y. (2007a). Empirical likelihood methods in econometrics: Theory and practice. In: *Advances in Economics and Econometrics: Ninth World Congress of the Econometric Society*, R. Blundell, W.K. Newey and T. Persson (eds.), pp 174-237. Cambridge University Press, New York.

62. Kitamura, Y. (2007b). Nonparametric likelihood: Efficiency and robustness. *The Japanese Economic Review*, **58**, 26-46.

63. Koopmans, L.H. (1974). *The Spectral Analysis of Time Series.* Academic Press, New York.

64. Kott, P.S. (1990). Robust small domain estimation using random effects modelling. *Survey Methodology*, **15**, 3-12.

65. Lazar, N.A. (2003). Bayesian empirical likelihood. *Biometrika*, **90**, 319-326.

66. Leonard, T. (1973). A Bayesian method for histograms. *Biometrika*, **60**, 297-308.

67. Little, R.J.A. (1982). Models for nonresponse in sample surveys. *Journal of the American Statistical Association*, **77**, 237-250

68. Lu, G. (2007). *Asymptotic theory for multiple-sample semiparpametric density ratio models and its application to mortality forecasting.* Ph.D. Dissertation, University of Maryland, College Park.

69. MacGibbon, B., and Tomberlin, T.J. (1989). Small area estimation of proportions via empirical Bayes techniques. *Survey Methodology*, **15**, 237-252.

70. Malec, D., Davis, W. W., and Cao, X. (1999). Model-based small area estimates of overweight prevalence using sample selection adjustment. *Statistics in Medicine*, **18**, 3189-3200.

71. Malec, D., Sedransk, J., Moriarity, C.L., and LeClere, F.B. (1997). Small area inference for binary variables in national health interview survey. *Journal of the American Statistical Association*, **92**, 815-826.

72. McGlynn, K.A., Sakoda, L.C., Rubertone, M.V., Sesterhenn, I.A., Lyu C., Graubard, B.I., Erickson, R.L. (2007). Body size, dairy consumption, puberty, and risk of testicular germ cell tumors. *American Journal of Epidemiology*, **165**, 355-363.

73. Mengersen, K.L., Pudlo, P. and Robert, C.P. (2013). Bayesian computation via empirical likelihood. *PNAS*, January 22, **110**, 1321-1326.

74. Monahan, J.F. and Boos, D.D. (1992). Proper likelihoods for Bayesian analysis. *Biometrika*, **79**, 271-278.

75. Müller P., Quintana, F.A., Jara, A. and Hanson, T. (2015). *Bayesian Nonparametric Data Analysis*. Springer, New York.

76. Nadaraya, E.A. (1964). On estimating regression. *Theory of Probability and its Applications*, **9**, 141-142.

77. O'Hagan, A. (1994). *Bayesian Inference, Volume 2B of Kendall's Advanced Theory of Statistics.* Arnold, London.

78. Orchard, T., and Woodbury, M.A. (1972). A missing information principle: theory and application, *Proceedings of the 6-th Berkeley Symposium on Mathematical Statistics and Probability*, **1**, 697-715.

79. Owen, A.B. (1988). Empirical likelihood ratio confidence intervals for a single functional. *Biometrika*, **75**, 237-249.

80. Owen, A.B. (2001). *Empirical Likelihood.* Chapman & Hall CRC, Boca Raton.

81. Pan, L. (2016). *Semiparametric methods in the estimation of tail probabilities and extreme quantiles.* Ph.D. Dissertation, University of Maryland, College Park.

82. Parzen, E. (1962). On estimation of a probability density function and mode. *Annals of Mathematical Statistics*, **33**, 1065-1076.

83. Parzen, E. (2004). Quantile probability and statistical data modeling. *Statistical Science*, **19**, 652-662.

84. Patil, G.P. and Rao, C.R. (1977). The weighted distributions: A survey of their applications. *Applications of Statistics*, Ed. P.R. Krishnaiah, 383-405. North-Holland, Amsterdam.

85. Patil, G.P., Rao, C.R. and Ratnaparkhi, M.V. (1986). On discrete weighted distributions and their use in model choice for observed data. *Communications in Statistics–Theory and Methods*, **15**, 907-918.

86. Pell, J.P. and Cobbe, S.M. (1999). Seasonal variations in coronary heart disease. *Q J Med*, **92**, 689-696.

87. Pfeffermann, D. (2013), New important developments in small area estimation. *Statistical Science*, **28**, 40-68.

88. Pfeffermann, D., Krieger, A. M., and Rinott, Y. (1998). Parametric distributions of complex survey data under informative probability sampling. *Statistica Sinica*, **8**, 1087-1114.

89. Pfeffermann, D., and Sverchkov, M. (1999). Parametric and semi-parametric estimation of regression models fitted to survey data. *Sankhyā*, Ser. B, **61**, 166-186.

90. Pfeffermann, D., and Sverchkov, M. (2007). Small-area estimation under informative probability sampling of areas and within selected areas. *Journal of the American Statistical Association*, **102**, 1427-1439.

91. Phadia, E.G. (2013). *Prior Processes and Their Applications: Non-parametric Bayesian Estimation.* Springer.

92. Phue, J.N., Kedem, B., Jaluria, P. and Shiloach, J. (2006). Evaluating microarrays using a semiparametric approach: Application to the central carbon metabolism of *Escherichia coli* BL21 and JM109. *GENOMICS*, **89**, 300-305.

93. Porter, A.T., Holan, S.H. and Wikle, C.K. (2015). Bayesian semiparametric hierarchical empirical likelihood. *Journal of Statistical Planning and Inference*, **165**, 78-90.

94. Prasad, N.G.N., and Rao, J.N.K. (1999). On robust small area estimation using a simple random effects model. *Survey Methodology*, **25**, 67-72.

95. Prentice, R.L. and Pyke, R. (1979). Logistic disease incidence models and case-control studies. *Biometrika*, **66**, 403-411.

96. Qi, Y. (2002). *Classification of microarray data.* M.A. Thesis, University of Maryland, College Park.

97. Qin, J. (1998). Inferences for case-control and semiparametric two-sample density ratio models. *Biometrika*, **85**, 619-630.

98. Qin, J. and Lawless, J. (1994). Empirical likelihood and general estimating equations. *Annals of Statistics*, **22**, 300-325.

99. Qin, J. and Zhang, B. (1997). A goodness of fit test for logistic regression models based on case-control data. *Biometrika*, **84**, 609-618.

100. Qin, J. and Zhang, B. (2005). Density estimation under a two-sample semiparametric model. *Nonparametric Statistics*, **1**, 665-683.

101. Ragusa, G. (2014). Approximate Bayesian inference in models defined through estimating equations. In: *Bayesian Inference in the Social Sciences*, I. Jeliazkov and X.-S. Yang (eds.), pp 265-290. Arnold, London.

102. Rao, C.R. (1965). On discrete distributions arising out of methods of ascertainment. In: *Classical and Contagious Discrete Distributions*, G.P. Patil, ed., pp. 320-332. Pergamon Press and Statistical Publishing Society, Calcutta.

103. Rao, J.N.K., and Choudhry, G.H. (1995). Small area estimation: Overview and empirical study, in B.G. Cox, D.A. Binder, B.N. Chinnappa, A. Christianson, M.J. Colledge, and P.S. Kott (eds.), pp. 527-542. *Business Survey Methods*, Wiley, New York.

104. Rao, J.N.K. and Molina, I. (2015). *Small Area Estimation*, 2nd Edition. Wiley, Hoboken.

105. Ren, J.-J. (2008). Weighted empirical likelihood in some two-sample semiparametric models with various types of censored data. *The Annals of Statistics*, **36**, 147-166.

106. Riddless, M., Kim, J.-K., and Im, J. (2016). A propensity-score-adjustment method for nonignorable nonresponse, *Journal of Survey Statistics and Methodology*, **4**, 215-245.

107. Robert, C.P. and Casella, G. (2004). *Monte Carlo Statistical Methods*, 2nd. Edition. Springer.

108. Rosenblatt, M. (1956). Remarks on some nonparametric estimates of a density function. *The Annals of Mathematical Statistics*, **27**, 832-837.

109. Rubin, D.B. (1976). Inference and missing data, *Biometrika*. **63**, 581-590.

110. Rubin, D.B. (1981). The Bayesian bootstrap. *The Annals of Statistics*. **9**, 130-134.

111. Sarndal C.E., and Swensson B. (1987). A general view of estimation for two fases of selection with applications to two-phase sampling and nonresponse. *International Statistical Review*, **55**, 279-294.

112. Schennach, S.M. (2005). Bayesian exponentially tilted empirical likelihood. *Biometrika*, **92**, 31-46.

113. Schick, A. and and Wefelmeyer, W. (2008). Some developments in semiparametric statistics. *Journal of Statistical Theory and Practice*, **2**, 475-491.

114. Sen, P.K. and Singer, J.M. (1993). *Large Sample Methods in Statistics. An Introduction with Applications*. Chapman & Hall, New York.

115. Shao, J. (2003). *Mathematical Statistics*, 2nd Ed. Springer, New York.

116. Shumway, R.H., Azari, A.S. and Pawitan, Y. (1988). Modeling mortality fluctuations in Los Angeles as functions of pollution and weather effects. *Environmental Research*, **45**, 224-241.

117. Silverman, B.W. (1986). *Density Estimation for Statistics and Data Analysis*. Chapman & Hall/CRC Monographs on Statistics & Applied Probability, New York.

118. Sverchkov, M. (2008). A new approach to estimation of response probabilities when missing data are not missing at random. *2008 JSM Meetings, Proceedings of the Section on Survey Methods Research*, 867-874.

119. Sverchkov, M. (2013). Is it MAR or NMAR? *2013 JSM Meetings, Proceedings of the Section on Survey Methods Research*, 2306-2311

120. Sverchkov, M., and Pfeffermann, D. (2004). Prediction of finite population totals based on the sample distribution. *Survey Methodology*, **30**, 79-92.

121. Tan, Z. (2009). A note on profile likelihood for exponential tilt mixture models. *Biometrika*, **96**, 229-236.

122. Thas, O. (2010). *Comparing Distributions*. Springer.

123. Thomas, D.R. and Grunkemeier, G.L. (1975). Confidence interval estimation of survival probabilities for censored data. *Journal of the American Statistical Association*, **70**, 865-871.

124. Vardi, Y. (1982). Nonparametric estimation in the presence of length bias. *Annals of Statistics*, **10**, 616-20.

125. Vardi, Y. (1985). Empirical distribution in selection bias models. *Annals of Statistics*, **13**, 178-203.

126. Verret, F, Rao, J.N.K. and Hidiroglou, M.A. (2015). Model-based small area estimation under informative sampling. *Survey Methodology*, **41**, 333-347.

127. Voulgaraki, A., Wei, R., and Kedem, B. (2015). Estimation of death rates in U.S. states with small subpopulations. *Statistics in Medicine*, **34**, 1940-1952

128. Voulgaraki, A., Kedem, B., and Graubard, B.I. (2012). Semiparametric regression in testicular germ cell data. *The Annals of Applied Statistics*, **6**, 1185-1208.

129. Wand, M.P., and Jones, M.C. (1993). Comparison of smoothing parameterizations in bivariate kernel density estimation. *Journal of the American Statistical Association*, **88**, 520-528.

130. Wald, A. (1944). On a statistical problem arising in the classification of an individual into one of two groups. *Annals of Mathematical Statistics*, **15**, 145-162.

131. Watson, G.S. (1964). Smooth regression analysis. *Sankhya* A, **26**, 359-372.

132. Wen, S. (2007). *Semi-parametric cluster detection.* Ph.D. Dissertation, University of Maryland, College Park.

133. Wen, S. and Kedem, B. (2009). A semiparametric cluster detection method - a comprehensive power comparison with Kulldorff's method. *International Journal of Health Geographics*, 8:73 (31 December 2009).

134. White, H. (1982). Maximum likelihood estimation of misspecified models. *Econometrica*, **50**, 1-25.

135. Yang, Y. and He, X. (2012). Bayesian empirical likelihood for quantile regression. *The Annals of Statistics*, **40**, 1102-1131.

136. Zhang, B. (2000a). A goodness of fit test for multiplicative-intercept risk models based on case-control data. *Statistica Sinica*, **10**, 839-866.

137. Zhang, B. (2000b). Quantile estimation under a two-sample semi-parametric model. *Bernoulli*, **6**, 491-511.

138. Zhang, B. (2001). An information matrix test for logistic regression models based on case-control data. *Biometrika*, **88**, 921-932.

139. Zhou, W. (2013). *Out-of-sample fusion.* Ph.D. Dissertation, University of Maryland, College Park.

# Index

www.ingramcontent.com/pod-product-compliance
Lightning Source LLC
Chambersburg PA
CBHW081516190326
41458CB00015B/5383